NO RETURN TICKET

NICHOLAS RETY

Copyright © 2022 by Nicholas Rety.

ISBN 978-1-956896-81-7 (softcover)
ISBN 978-1-956896-82-4 (ebook)

All rights reserved. No part of this book may be reproduced or transmitted in any form or by any means, electronic or mechanical, including photocopying, recording, or by any information storage and retrieval system without express written permission from the author, except in the case of brief quotations embodied in critical reviews and certain other noncommercial uses permitted by copyright law.

Printed in the United States of America.

Book Vine Press
2516 Highland Dr.
Palatine, IL 60067

to Kati, Sári, Zsuzsi
and Éva

CONTENTS

FIELD OF FIRE 13
The siege of Budapest, 1944-45, at close quarters

ALTAR EGO 63
Humiliation in public

NIGHTWATCH 73
Catching a thief in the wartime blackout

IN GOES A BOY – OUT COMES A MAN 79
How I conquered my fear of the dark

ONE LAST LOOK 85
How I conquered my fear of the dark

EYE ON THE BALL 89
Encounter with the secret police

THE TRUSTED MAN 93
My neighbour, the informer

NIGHT CROSSING 97
A journey to remember

HOLIDAY FOR GREED 101
Uncontrolled pandemic

THE CRUCIBLE 107
A life-changing, life-saving journey

OLD SALT 121
My old friend, the sailor

MY HOME, MY CASTLE 127
Tough student days in London

MARMALADE WARS 139
Foot soldier in the marmalade war

JELLIED EELS FOR A PRINCE 145
Death of a simple man

THE QUEUE – JUMPER 147
A short-cut to heaven

THE VIGIL 155
My first emergency

OFFICIAL SECRETS 163
Private lives

CENTRE OF THE WORLD 171
A riddle solved after 60 years

BERNIE'S WAR 177
Murmansk convoy memories

LIBERTÉ, ÉGALITÉ, FRATERNITÉ 183
Did those heads roll for nothing?

THE ART OF WARFARE 187
The ancient game has it.....

MIRROR WITH NO IMAGE 193
Can we judge ourselves?

CURTAIN CALL 201
Return behind the Iron Curtain

GHOSTS ON A STAIRCASE 207
Echoes of the dead

SOLDIERS OF THE QUEEN 209
Voice from the lower ranks

DIRTY MONEY 215
Flirting with trouble

JANUARY RIVER 227
Rio de Janeiro adventure

SANDCASTLES 241
The past is elusive

LOVE BETWEEN TWO ISLANDS 251
Living an illusion

HIDE AND SEEK 263
Death of a little girl

SPRING CLEANING 269
Visit to a rattlesnake den

POSTMARK: TAKLAMAKAN 275
A letter, not delivered

LA DAMA BLANCA - A PILGRIMAGE 281
Pilgrimage to the ship where my friend died

HOW DO YOU SAY IT IN GERMAN? 289
The journey, not the discovery

STRIKE THE IRON 295
Time does not wait

CLOUDS FOR COMPANY 299
Learning to fly at fifty-six

SILENT KILLER 331
An essay on prejudice

TO AUSCHWITZ – AND BACK 339
Alone in the silence

FAREWELL JANOS, MY BROTHER 351
My brother, my friend

CHARITY BEGINS – AND ENDS 357
A new industry

ORGANIC SOLUTIONS 361
Who is in control?

LETTER FROM HONG KONG 371
A city to remember

LETTER FROM BALI 377
People and customs of the island

VILLAGE ON THE MOUNTAIN 391
In search of the real Bali

The Pirates and Beggars of Lake Batur 397
Strange life, stranger death
(article from The Bali Times)

When You Have Nothing, Yet Have It All 403
A life untouched by greed
(article from The Bali Times)

BALI MORNING 407
The birth of a day in Bali
(article from The Bali Times)

EPILOGUE 411

Our footprints in the sand are washed away by the ebb and flow of tides, leaving no trace of our passage. We must strive to leave some record of our journey through life. Merely passing on our genes is a biological inevitability which says nothing of ourselves. Our thoughts, ideas, words, and the images they convey should endure to show that we passed this way. Maybe a hint of wisdom from our lessons in life will light a candle of understanding held by an unseen hand.

FIELD OF FIRE

FIELD OF FIRE

City Under Siege

PRELUDE TO BATTLE

The events I am about to describe I would rather forget. They took place more than half a century ago. They should have found their way into the litter-box of my memory by now, along with the flotsam of a long life. Curiously, they won't go away. Unwanted baggage, they stay with me. Even if I try to lose them deliberately, shut them out, they turn up at the doorstep of my consciousness time and again. I even looked in the mirror, expecting to see "Budapest: 1944-45" emblazoned on my forehead but found no outward sign. Yet I know that what I am going to say is engraved somewhere within me and will accompany me, like a shadow, for the rest of my days.

The notion that I was somehow enriched by these events is called into question by the circumstances in which they took place. Yet I cannot deny that the lessons I learned from their passing have shaped my thinking for life.

I recognized that life as a force is more difficult to extinguish than I had thought. I came to realize that, no matter how tenuous your hold on life, it is worth hanging on. Most of the time things will work out in a way to leave you with some choice at least.

I now know that when you are in uncharted territory, when all support is gone, when there is no script to tell you how to act, which way to go, you must trust your instincts. The unrecognized, often unexplored or even considered inner self has resources in plenty to see you through events to which you might otherwise have surrendered.

At the end, whether you emerge as a survivor or victim is up to you.

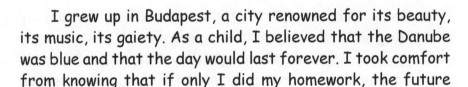

I grew up in Budapest, a city renowned for its beauty, its music, its gaiety. As a child, I believed that the Danube was blue and that the day would last forever. I took comfort from knowing that if only I did my homework, the future would take care of itself.

A growing child needs a sense of security. This is found in a stable home where there is no threat, where events take place in a predictable pattern, where there is love, food and warmth.

Such insecurities as I did have, all related to our lack of money. Luxuries, such as family holidays, we did not know. We travelled by streetcar and only occasionally did I enjoy the thrill of a ride on the bus. The end of the streetcar line, for all intents and purposes, was the end of the world.

My first journey on a train is memorable for the fact that I sat on the hard, wooden bench of a third-class compartment on the way and when the train stopped, my very best friend, alighting from the adjacent plush second-class carriage, pretended not to know me.

As a boy grows older he needs to widen his perimeter. I longed to own a bicycle but rode one only in my daydreams. As for a family car, I knew we would never own one, so the folding window pane became my windshield and I took it for long rides to places that existed only in my imagination.

At dinners with the greater family present the children always sat at the lower end of the table. From the conversation of the adults we learned that a war was going on somewhere but did not know what that meant. In time, such talk became more serious and for the first time we had an approaching sense of threat. Still, it was all in the abstract.

The door burst open in 1944 and the threat was abstract no more. Bombs rained from the sky. I very quickly understood the fear of imminent death and my inability to do anything about it. German soldiers were seen everywhere and were deadly serious about taking control. Threatening posters appeared, people, some our friends, vanished overnight. Martial law was declared. Bursts of gunfire were heard in the night. The light of morning often revealed rows of dead bodies in a nearby park. People began to talk in whispers as it was no longer safe to speak out loud.

Then, one afternoon, I heard the rumble of a distant thunderstorm. Unlike other storms, this one did not end but continued into the night. The lightning bolts lit up the sky in the east for hours on end. Slowly I realized that this was no thunderstorm at all: a great battle was being fought close enough to visit our senses. The war, the distant threat of dinner table talk a few years before, had reached the far outskirts of the city. The fury of the sound of guns filled me with awe.

I was almost fourteen.

BAPTISM OF FIRE

A watermelon is an innocuous fruit, almost comical when it grows to a certain size. As I was taking generous helpings from it one hot August afternoon in 1944, I did not know it would put my life at risk.

The village of Csömör (translates as "surfeit") lies some 16 kilometres outside Budapest to the northeast. We rented a room in a house made of mud bricks along Andrassy Street. The street led to an intersection with the main road leading into town and continued straight on as a tree-lined avenue through farmed fields. At the intersection stood a tavern, notable for two things. The owner had a pathological fear of bombing. He also had a lovely daughter, pursued, alas, by the son of a wealthy farmer.

We moved to Csömör to get away from the bombing, a good decision, as in the event our house received a direct hit and other severe damage besides. What we did not know was that the army also looked at the village as a haven and that every time there was an air raid in the night, they would drive a whole column of armoured cars from a depot some distance away to the shelter of the avenue of trees in line with our house.

Castle Hill in Buda, Royal Castle in foreground

In the late summer of 1944 the Russians took care of night bombing. You could tell by the sounds of the airplanes and by the bits and pieces such as bomb fragments and unexploded bombs they left behind.

There was a Russian air raid the night I consumed all that watermelon. This time, however, the planes flew quite low and close by and in no time the night was lit up by parachute flares. They turned the moonless darkness into broad daylight. Obviously they had signals from the ground as the flares were dropped right over their intended target: the armoured car column down the road.

At first they made several passes, dropping bombs. They made a lot of noise but their aim was not good. One of the bombs dropped close to our house but failed to explode -- we found it the next day in the field with its fin still above ground.

While the bombs were dropping the watermelon enforced its inevitable effect on my bladder and I had to go to the bathroom urgently. The problem was that the flares still lit up the village and the toilet was an outhouse down the garden. It was unwise to attempt to go outside. The approaching drone of aircraft engines heralded another attack. By now I was very uncomfortable. I stood on the steps under an overhang, looking at the outhouse which looked so desirable, even beautiful, in the eerie daylight created by the flares.

The planes approached and I knew I could not leave the house. Just then I heard a machine gun open up and I saw the dust thrown up by the bullets just outside the steps. I remember my grandmother's voice urging me to join her under the table, which she quite correctly identified as the only refuge in the event the roof collapsed. There were three planes going around in circles, strafing, now that they had discharged their bombs. I tried to figure whether there was enough time to make it to the outhouse and back between passes, but there was not. Because the armoured cars were in line with us, we also got the strafing. The airmen found their target and were not letting go.

My distress turned into agony. I had to do it, but with the planes coming around in circles I could not risk going to the outhouse. The dust from the third plane's bullets had not yet settled when I was forced to do some watering from the bottom step. It did not keep the dust down for long.

As the Russians advanced, night bombing became more frequent. The army still continued to bring the armoured cars under the avenue of trees by the tavern. The owner then decided to build a bomb-proof air raid shelter. Work began at once. I watched the workmen dig deep into the ground, then place reinforcing beams. Entrance was made by a vertical ladder. Much to my pleasure we were invited to share the shelter and none too soon as there was an air raid alarm that very night. It took us a few minutes to get there in the pitch dark. The owner waited for us. A candle burned deep in the shelter where it could not be seen from the outside. Descending the steps felt like entering one's own grave. In a way, that almost came to be.

The door was closed and we sat by the uncertain light of the candle. The air was stuffy, the space small. I even wondered if the candle would go out as we ran out of oxygen. The taverner was thanked and congratulated on his foresight to build so secure a shelter. Barring a direct hit, it should stand up to anything.

We could soon hear the planes. With the front getting closer they took no time to reach us. They were low again and very close -- the engines roared overhead and suddenly the ground beneath our feet leapt as the bombs hit. Phew! That was very close. More bombs followed, but none closer. There was no strafing, the raid took only a few minutes. The "all clear" soon sounded. Emerging from the shelter we found four very shallow craters close together some thirty yards off. Large trees in the immediate vicinity were blown away by the blast. The bombs, obviously chained together, were the kind that explode above the surface and do damage by blast and splinters. Luckily for us, they did not challenge the shelter.

It also became obvious that the tavern and its shelter were literally at the bull's eye of the bombing target as long as the armoured cars continued to return. Besides, there was no doubt that someone was using very effective signals to guide the planes to their target. Mud-bricks or not, it was safer in our little house two hundred yards away.

As the ground battle came within earshot we moved back into the city.

The bombing continued both day and night.

A terrible game of chance, sitting under the bombs. If you are anywhere close to the target of the day, you hear the planes coming closer and closer, and when you begin to think that they will go right by, you hear the bombs. There is no escape. You don't know where they will hit so there is no point in running. They fall faster than you can run, so you just wait. You hear them getting louder and louder as they whistle through the air. You try to shrink into something insubstantial as if you could make yourself into a smaller target. You know you may die in less than a minute, perhaps in seconds. As the whistle becomes a roar overhead there is no time to think. Suddenly, everyone is alone. There is no time to pray.

When the bombs hit, the ground leaps underfoot and the sound is deafening. In the fitful urgency of the moment all rational thought ceases -- you simply die, only to find you are alive again until you die once more as the next wave hits. Then, suddenly all is quiet again. The ground is still. In the air the sound of the airplanes quickly fades. The whole thing lasted a minute, maybe two. In the shelter people sit upright again and finally someone speaks. When the siren signals "all clear" it will be time to go upstairs until the sirens signal another attack, sometimes only minutes later. Civilians living under such attack are helpless. It may be your turn or someone else's turn next time. There are no winners, only losers.

The Russian ring around the city was closing fast. It was clear that when it finally closed, a savage battle would follow. Some people, fearing the prospect of a war on their doorstep, some fearing the arrival of the Russians, fled to the west, prolonging their war and departing for an uncertain future. I remember hearing my old uncle recite the words of an old Hungarian anthem to a man who was in two minds about going:

> *"should Fate's hand bless you*
> *or strike you down, here is where you live*
> *and here is where you die."*

The rapid advance of the front cut off supplies of food. The artillery shells were finding their way into the city. This made it hazardous to move about. Soon the streetcars and buses stopped running. With martial law firmly in place, one could not venture far as anyone seen on the street after 5 p.m. was to be shot on sight.

People meeting in the street no longer stopped to talk. Where, before, they would stop and chat about family and friends, they now exchanged curt greetings and continued on their way. Survival became everyone's main concern now that shells came from every direction without warning.

There began a process of increasing isolation of one individual from another. During the bombing people seemed to withdraw into themselves to deal with the recurring threat of death in their own way. At least with the bombs the wailing sirens gave warning. Not so with the shells. They kept coming all day. Social contact, visiting and sharing meals with friends all ceased, not only because of the danger of moving about or the reduced perimeter imposed by martial law but because there was no longer enough food to share with friends. This animal way of protecting food supply was acutely felt in a society where sharing food and drink with friends had been a way of life. As food became more and more scarce the sense of isolation grew. Those without food were too proud or too embarrassed to admit their plight. Those who had food stayed away from others lest they were embarrassed into giving food away.

Our supplies of food were pitiful and for the first time in my life I came to know hunger. I was hungry all day, even after our meals. The concept of sharing became a reality as we took small helpings to make sure everyone had fair share.

THE RING CLOSES

On Christmas Eve 1944 we sat down to a spartan dinner in the kitchen. With the windows in the other rooms blown away in an air raid some months before, the kitchen was the only place where we could keep warm. There was no Christmas tree and we had no gifts. The sound of carols was replaced by the whistle of artillery shells. The water in our glasses rippled with each explosion. We were anxious about what was to come and helpless to do anything about it. Talk about Christmas Eve dinners of years past gave us comfort of sorts but the simple fare on the table brought us back to the reality of the moment. The fear of the unknown visited all of us. It came to stay.

There was a knock on the door. Our neighbour came to tell us that the Russian ring around the city was now complete. The real battle was about to begin.

It was time to move down to the air raid shelter. Even though we had been living and sleeping in an apartment without windows, at the mercy of the elements, the prospect of staying in a large, musty cellar with a lot of strangers was unappealing. Like it or not, the cellar was to be our home for the next two months.

On Christmas Day the shelling was heavy but I went to church. A young priest was saying his first Mass. The church was full. Many in the congregation must have believed that it was their last visit there. During the service an artillery shell hit the building, sending debris and glass flying. The service continued. When it ended I stepped out into the crisp, sunlit morning and heard the guns fill the air with menacing sound. In the street I saw a group of heavily armed German soldiers, dressed in camouflage uniforms, marching in step, singing defiantly:

"......rechts und links marschieren g'rade an"

The battle for Budapest lasted 59 days and was particularly savage. The story got around that, when the city was surrounded on Christmas Eve, the Russians sent a Captain Ostapenko across the lines with a white flag. He carried a message asking for the surrender of the German garrison defending the city. The Germans refused to surrender. As Captain Ostapenko walked back to the Russian lines the Germans shot and killed him. The veracity of this story was later questioned but of the savagery of the battle there was no doubt.

By day the Stormovik fighter bombers ruled the sky. I watched them from the window of our apartment as they circled the Citadel, strafing and bombing, hours on end. Their attention was caught by anything that moved on the main roads. One day I was walking with my mother along the road when aerial activity was light. Suddenly, a machine gun opened up behind us. My mother pushed me into a doorway and as she lunged after me the sidewalk was peppered with machine gun bullets, stirring up dust. Only when the shots were already well past us did we hear the Stormovik's engine come alive with a roar. As we looked up, there he was, climbing again. Gliding in on the target with the engine idling they achieved complete surprise.

One day, about sunset, a crippled Russian plane flew over the city. It was flying very low, obviously trying to make it back to the Russian lines. Every anti-aircraft gun, machine gun, submachine gun, rifle and pistol was firing, trying to bring it down. The sky was alive with tracers from all directions. You just had to feel for the pilot. It seemed inevitable that he would be hit. He flew on, crawling on air, it seemed. At length he was safe from the guns of the ringed city and I hoped he made a safe landing.

Attacks from the sky sometimes worked to our advantage. Word got around one morning that a horse had been shot and killed nearby and a butcher was cutting it up for meat. I have no taste for horse-flesh but, hungry, I went along. By this time the Germans had built tank traps to slow down the Russian advance. Some were five-foot high concrete pyramids strewn about the width of the road. Others were fifteen-foot deep V- shaped excavations wall to wall across the street. One of these V-shaped traps was located right outside our butcher's shop. Down there in the "V" stood our butcher, apron donned and knife in hand, cutting chunks of meat for people standing in line above. I got my piece. There was no wrapping or bag to be had, so I carried the raw, dripping horsemeat home in my bare hand. By then we had had no meat for some three months, so it was a windfall of sorts.

In a way, every day had its high point. One morning a solemn German came into the air raid shelter and told us that the nearby distillery would be blown up soon to stop the Russians from getting at the liquor. We were free to help ourselves to anything we wanted during the next two hours. I quickly picked up a bucket and was off. Outside it was bitterly cold. The distillery was about 200 yards away along the main road. The Russians were advancing and were quite close by but a bend in the road protected us from small arms fire. Mortar activity that day was light. The sight was surrealistic. Scattered in the snow lay drunken bodies hugging buckets, oblivious to the firing. They were taking time out of war and they were happy. I don't know how many froze to death but if they did, they did not seem to mind.

I entered the distillery through the big wooden gate, following footprints in the snow to a big door inside. The door led to a large hall, much like a church with steps leading to a "souterrain" below, with big vats on both sides. Another set of steps led to a mezzanine above. The vats on the mezzanine were busily attended but down below in the souterrain there was less activity. Unfortunately, the taps to some vats had been left open and they were gushing out their contents in torrents. The souterrain was filling up with liquor from the floor up. Still the vats were waiting and I wanted to get out as quickly as I could (the Russians were just past the bend in the road), so I waded in, knee-deep in liquor, to the nearest vat and filled my bucket with some sweet stuff. I took a long swig and it tasted good. By then there was no such thing as breakfast, so it was my first "meal" of the day. As I waded out, bucket in hand, someone offered me a drink from his bucket, so I took a big gulp from that. His tasted better than mine but now it was time to go. Once I got up onto dry land, my wet feet began to freeze in the cold. I hurried along as fast as I could, trying not to spill my treasure. Now and then I would stop for a drink on the way. I don't recall hearing any shooting on the way back.

When I reached the shelter my head was spinning. I could smell the sickly aroma of liquor on my clothes. I lay down on my bunk. The last thing I remember is a circle of faces staring down at me. I tried to speak but could not. Blissful sleep followed. No shooting, no dreams, no war. I woke up the next day.

The siege of the city seemed endless. The encircled Germans were doomed and put up terrific resistance. All the while we lived underground, in a large musty cellar, everyone thrown together in one large damp, ill-lit space. We all had makeshift beds, ours were immediately inside the door. We awoke, waited, talked, rested and occasionally ate, down there. We had very little food and ate mostly boiled beans, just once a day. The battle promised to go on and on, so we had to conserve what we had.

Not so the rich merchants of the building. At mealtimes they all disappeared upstairs. The enticing smell of good food emanating from their apartments had me go upstairs to the courtyard at such times just to smell what I could not eat. They never shared their food, the question of eating was not even discussed. The richest of the merchants was a man named Temes. His kitchen produced the best smells.

The shelling was so bad that we could not use the bathrooms and toilets in our apartment, so for a while, everyone had to use a toilet on the ground floor. Soon the water supply ceased to flow altogether and the toilet overflowed and became unserviceable. It was buckets from then on. When the firing was particularly severe, we just had to urinate in the cellar, with everyone present. I still remember how it bothered me to hear my mother use the bucket during a lull in the shooting. By then, however, no one seemed to care. Most of us were hungry all the time and there was much doubt as to whether we would survive much longer.

When the water supply failed we had to go to a nearby flour mill to fill our buckets from a well. This became a hazardous pastime as the Russians neared the bend in the road. As soon as they had line of sight we had to give up our water trips. They were almost there one morning when

I made the trip with two buckets. There was heavy firing and shells and mortar bombs were hitting within sight as we lined up for water. In the bitter cold the sound was more intense than ever. Everyone was anxious to get back underground and the hazards of being exposed in the open made the waiting seem exasperatingly long.

A middle-aged, tall German soldier appeared. He awaited his turn but was as anxious about the firing as the rest of us. Soon he pulled out a pistol with a very long barrel and started firing in the air, as if to pump up his own morale.

My buckets filled, I headed back to the shelter at last. A quick look to the left as I entered the road but I did not see any Russians so I hurried back to the house. The buckets felt heavy and I did not want to spill any water as it was now a precious commodity. Further trips to the well were unlikely for a few days. On entering the building I had to get through the central courtyard to gain the entrance to the spiral stairway leading to the shelter at the far end. I was half-way through when I felt my eardrums being pushed in, then pulled out, felt a blast around me and saw dust flying. A mortar bomb dropped on the opposite side of the courtyard, some 20 feet from me. It hit the stone walkway, left just a shallow mark on the stone and covered the wall with splinters. There were splinters on the wall on my side, too, but I was untouched. There was no time to react, so I just carried the water down to the shelter and contemplated my good luck.

There were other hazards, too. As the regular water supply failed, we could no longer bath or even wash. One day I began to itch badly around the waist. My mother made me undress completely and ran a hot iron through my clothes to kill the lice. Most of our drinking water on that day went towards a thorough wash over a small basin. I did not get lice again.

There was no fresh air. I longed to leave the shelter and breathe clean air. Of the latter there was none. After the heavy bombing of the last nine months the air was heavy with the smell of burning and smoke. Every breath was a reminder of destruction and death. Worse, our building received severe bomb damage months before with a twenty-foot crater just in front and a direct hit on the side-street side. The bomb penetrated the sewer beneath and the rubble was overrun by rats. I amused myself by throwing rocks at them. I never hit one.

On January 13 the sound of small arms fire became intense. The Russians had reached the bend in the road and were approaching our house. Firing was severe all day. In the evening a group of seven or eight Germans led by a major, came into the shelter to have their meal, brought up in a big can by their field kitchen. The meal was one of beans, laced with meat flavouring, but contained no meat. The major sat by himself and was polite and laconic. The rest of the soldiers were very quiet. One of them was only seventeen. When they finished eating they started cleaning their weapons.

The seventeen-year old took his pistol apart and oiled it carefully. He said he was going to shoot himself with it, rather than be captured. They soon left.

On January 14 the front reached our house. Small arms fire was endless all day. We stayed underground, not knowing what to expect. No one spoke. It was disturbing to think that somewhere upstairs people were actually trying to kill each other. We were helpless, so we stayed put. In the afternoon, there was loud commotion upstairs. We heard the trampling of many feet, and excited shouting. Three of us went upstairs to investigate. We found some forty terrified Germans, armed to the teeth, in the courtyard, trying to find a way out of the house. They had just rushed in through the front door and told us the Russians were across the street. The last thing we wanted was a pitched battle in the building so we told them to get out through the side street. As they left they asked if we could do anything for the wounded man just inside the door. I went to take a look. The front door was closed. Just inside it lay the German on his back. His arms and legs were spread out, his rifle lay beside him. I took a closer look. His mouth and eyes were open. Between his eyes, right at the root of his nose, in the very centre, there was a bullet hole. He was dead. I marvelled at the symmetry of it. Obviously, as the soldiers rushed in and slammed the big door behind them he chose to take a look at the pursuers through the grill. It then dawned on me that whoever shot him must be just outside that door, so I made a quick retreat to the shelter. In the days that followed I retrieved his dog tag and hung it on a pipe in the shelter but it disappeared. I suppose he was listed as "missing in action."

Fighting was savage for twenty-four hours. We did not go above ground again. The worst of it was that we did not know what to expect, especially as the Russians must have seen all those Germans rush into the house, so they may think they were still there. If so, we could expect a noisy greeting, maybe a hand grenade, before we saw the first Russian.

On January 15 the shooting stopped abruptly. The silence was full of questions. All conversation ceased.

We waited. We expected something to happen, anything just so that we could know we were safe at last. Instead, the silence endured and, as it grew longer, so our anxiety mounted. The fear of not knowing is the worst fear of all. You cannot run from it, as you don't know where to run. You cannot fight it for it has no shape or form. There is no choice. You just wait.

It must have been four in the afternoon when we heard footsteps. Whatever was going to happen would happen now. The strange thing was that we heard only one man walking, right above us. A few steps now, then he would stop, then some more steps and he would stop again. My father spoke Russian, so he decided to go and take a look. I followed. We inched our way up the circular stairs just enough to catch a glimpse of the courtyard. There was no one there. We waited. All of a sudden a Russian soldier appeared with a fur hat on, looking a little anxious. He had a pistol in his hand. He did not see us. My father called out in Russian and the soldier spun around. He was very young. As my father continued to talk to him a big smile came over the soldier's face. "Any Germans down there?" he asked. "Only civilians" came the answer. He put the pistol back in the holster and I met my first Russian soldier of the war, Vladimir Vasilyevich Kutyminski.

Our relief was immense. He came down into the shelter and put us at ease. He was relieved, too, as he had expected to find Germans in the house. He told us we had nothing to fear and soon left.

Before sunset we ventured upstairs again. It was good to see daylight at last. The dead German still lay there but the door was now open and his rifle was gone. His eyes were still open. I took a look out of the front door and I saw women soldiers carrying mine detectors along the street. They had large bosoms, pulled very high under their tunics. The men were mostly Mongolian. They were moving up big guns and pointed them at the Citadel. The three Danube bridges we normally could see from the door were all in the water. The Germans blew them up in their retreat. The railway bridge to the south was the first to go on December 29, in what seemed like a terrible earthquake. The Horthy Bridge, named after the Regent of Hungary, went down on January 9. This bridge being closer, the noise and tremor were much worse. By this time we knew what was happening. We felt a terrible anger at the Germans for destroying everything in the wake of their retreat. In fact we knew months ahead that this would happen when three sections of the Margaret Bridge, connecting Pest to Margaret Island were accidentally blown up while carrying heavy afternoon traffic. The incident received hardly any publicity and no official comment.

In the evening of January 15 I stood at the window, listening to strange new songs of the Russian soldiers nearby. They were unlike any songs I had heard before. One soldier would start and the others join in. I liked their music.

On January 16 a signals unit took up residence in the house, led by a Cossack officer. He wore a Persian lamb fur hat with a cross on top. He came down to make sure we were alright. The women found him very handsome. He had good manners. We wanted him to stay for the rest of the war. In the evening he returned and asked if we had a guitar. My mother gave him her guitar. He played and sang beautiful songs for an hour or so, then he left. My mother wanted him to take the guitar but he would not. He said they were moving on in the morning. Looters took the guitar two days later.

January 17 was cold with no sun. I was standing in the road, watching the goings-on when three drunken soldiers appeared. They tossed a nearly empty bottle in my hand and invited me to drink to Stalin. They looked bad and the slime at the bottom of the bottle was revolting. I refused to drink. At that point they turned angry and insisted that I drank. I lifted the bottle to my mouth and pretended to drink. They were not fooled. "To Stalin," said one, pulling out his pistol. I drank. The first thing in my mouth was the slime. I quickly swallowed it and tossed the bottle back to them. They turned and went.

There were rumours of rape and plunder in the wake of an advancing army and of reprisals against men who would try to stop it. So far no one bothered us. We felt safe.

In the evening of January 17 we were sitting in the shelter in the dim light of a hurricane lamp when the door opened and three Russians walked in. They looked around, then sat down at the table and pulled out some cards. The biggest of the three, Nikolai from Vladivostok, as we learned, was drunk. He sweated profusely. They motioned to me to join them in a game of blackjack. I had no money, I explained in sign language, so they gave me some "occupation money," actually printed for the Russian Army in our currency. It was worthless. The game began. Everyone was quiet. It was hard to see in the poor light. After a few games Nikolai got up and left the table. The rest of us continued playing.

It was then that I heard a whispered "No!" from the darkness behind me. I turned to find Nikolai dragging my mother toward the open door. My mother, fearing I would get hurt if I intervened, tried to pull away but her strength was no match for the Russian's. I jumped up, tossed the cards on the table, took hold of my mother's arm and gave

Nikolai an almighty shove. He was enraged. He let go of my mother, got hold of me with both hands and began to yell, spitting anger, alcohol and saliva right in my face. He pushed me against the wall and picked up his submachine gun, still yelling. As he cocked his weapon the other two soldiers began to shout at him. He half-turned to argue with them. When he looked away I knew it was my only chance. I made a leap for the door, pushing my mother before me. We ran, stumbled and scrambled in the darkness toward the small shed where we kept wood and coal for the winter. We moved into the shed to regain our breath. It was a poor hiding place but it was our only choice. It was pitch dark and very cold. We did not talk, in case we were being followed. We dared not move in case we knocked something over and gave our position away.

An hour passed. We had only thin clothes on and we were beginning to shiver. There was no source of heat; we could not stay there as we were. I decided to go back to the shelter for blankets. Maybe the Russians were gone. I told my mother to stay still till I returned.

I felt my way along the damp wall in the dark. The cellar had a dirt floor so I could move quietly. On the one side there were skylights to the courtyard, on the other, wood sheds. Once I turned into the long, dark corridor I could see the dim light issuing from the shelter through the open door, about a hundred feet away.

I stopped often and listened for sound. There was none. Maybe the Russians had gone. I inched up to the edge of the door and waited. The steel blast door which opened outward was open wide. The inward-opening wooden door was half open. There were voices inside but I could not make them out. I leaned over enough just to take a look with one eye.

Nikolai was standing with his back to the door about fifteen feet away. He had everyone lined up facing him and kept saying "DOKUMENT!" whereupon someone would produce a "document" which he would carefully examine and return until it was the next person's turn. The other two Russians were nowhere to be seen.

Our bunks were right by the door. I decided to move in, pick up a couple of blankets and move out. I was confident I could do it without his seeing me and was sure no one facing Nikolai would give any sign of recognition.

I moved into the doorway. One of our neighbours looked in my direction but he looked right through me. I tiptoed to the bunks, keeping my eyes on Nikolai. He was busy with the document in his hand, muttering in Russian. I reached for the first blanket. I pulled it off easily, without sound. I gathered it under one arm and went to work on the other. No problem. With blankets under both arms, I began to back out toward the door.

Everyone saw me by now but gave Nikolai no clue. I reached the doorway, just a step or two to go when the wooden door began to follow me with a creak as a blanket became caught in it. Nikolai spun around. He recognized me instantly. He started shouting and beckoned me to go up to him. He had a pistol in his right hand. "DOKUMENT!" he kept repeating. I then remembered that among my things I had a tramway pass, complete with photograph. I gave him my "dokument."

The photograph showed me wearing my school cap. He took one look and began to scream "PARTIZAN!............"PARTIZAN!", pushing the pistol barrel into my chest. He was in a rage and there was nothing I could do.

Just then one of the older women went up to him, put a hand on his arm and, in a way he must have understood, pleaded that I was no partisan. "PARTIZAN!..........PARTIZAN!" - he kept screaming and poked me with his gun.

To this day I do not know how I got out of there. I found my way back to the shed, my expedition having failed. I had no blankets. We were freezing. Nikolai knew that we were nearby. If he really thought I was a partisan he would surely look for me and kill me.

We huddled for warmth but still shivered uncontrollably. It was bitterly cold. Atop the grill to the street a few feet away we heard the Russian sentry slapping himself to keep warm. We could not risk making a sound.

The night seemed endless. I had a ghastly, gouging feeling in the pit of my stomach. The darkness gave my fear a new dimension. I thought I would never see daylight again. Daylight itself would seal our fate as Nikolai would have no trouble finding us then. We were cornered. There was nowhere to go, nothing to do but wait.

Time and again I tried to pray. I never got past the first few words of the Lord's Prayer. I was so cold that I could not harness my words together. "Our Father, who art...." I began again and again but I could no more control my thoughts than control my shivering. A dreadful sense of weakness came over me as my body heat ebbed away. I knew that I would die within hours, maybe minutes. There was no appeal, no redress, no hope.

I tried to reach God in a clumsy way through garbled words silently spoken. I received no response, no reassuring touch, no sign that all would be well. God did not know of our predicament or maybe he just did not care.

The pervading, numbing cold made it very difficult to prepare to die. With no comforting word from God, I found some solace in recalling hours I spent with my cousin in the heat of summers past, sitting at the top of an old apricot tree, enjoying the fruit, not a care in the world. Even when I tried to pray, I found myself back in the apricot tree again.

The sentry above was slapping himself in the cold. There was no shooting. We dared not move for fear of knocking something over in the dark. We breathed through the mouth, trying to avoid making any sound. I ceased hoping for daylight as it would only betray our hideout. I had to face it: there was no escape. If only I did not die shaking like a leaf!

Footsteps! The unmistakable sound of someone approaching made our hearts race with sudden fear. So this was it, in a pitch dark cellar, in dead of winter! I waited for the flash of the gun. That dreadful, sinking feeling hit me in the stomach again. The footsteps came closer, stopped, then seemed to move away. Quite quickly they returned and now our pursuer was a few feet away, outside the shed. He stood and waited. We held our breath.

"Miki" came a whisper from the dark. "Miki, where are you?"

"Over here", I whispered, still conscious of the sentry outside the skylight. I reached out and touched him with immense relief but I was all in. He led us back into the passage. The old concierge had been looking for us for some time. He told me Nikolai had led a woman out of the shelter and raped her. Nikolai was gone, all was quiet and it was safe to return. We wasted no time and soon we were thawing out under our blankets. We had been hiding for seven hours. It was now 3.30 in the morning.

We had not been in bed five minutes when the creaking door opened and in came Nikolai. He walked into the middle of the cellar and looked toward the woman he had raped earlier. As she stood up and walked to him with a look of resignation we pulled the blankets over our heads and wished we had died in the shed.

Zsuzsa Temes was a pretty girl of about twenty. She was the daughter of the rich merchant. Her mother feared greatly for her daughter's safety. Zsuzsa was an obvious target for rape. One day her mother asked me to rub Zsuzsa's face with dirt to make her less attractive. I had not touched a girl before and found the exercise most pleasant. I do not know where Zsuzsa was the night Nikolai visited but she certainly was not raped.

One evening, perhaps January 18 or 19, a friendly Russian came in, sat down and carried on the kind of good-natured conversation in which people engage despite being unable to speak each other's language. Everyone was amused. He was offered a cup of tea. While sipping it, his eyes strayed to Zsuzsa. He put down the cup, got up and with a smile on his face he walked over to Zsuzsa. He pulled her up by the hand. As he began to lead her away to Zsuzsa's protests the mother sprang to her feet and wedged herself between them, protesting firmly. The smile left the Russian's face but he did not let go of his grip. He was determined to take his prize.

When Mrs. Temes failed to make any impression on the Russian she suddenly turned to me and said "Miki, you can help! Give him your camera, so he will let Zsuzsa go!" I hesitated. True, I had a camera but by now everything had been taken from our apartment. The guitar was gone. The only clothes we owned were the ones we were wearing and the camera was the last thing of value I possessed. One day soon I could barter it for food., maybe even meat! Besides, the Temes family had lots of valuables to offer the soldier, they were rich and their apartment had not been looted.

Mrs. Temes pressed me. "Miki, the war will soon be over and when it is, I shall buy you anything you want, anything! Just give it to him!" I looked at Zsuzsa and saw fear in her eyes. "Here, take it!", I said and gave Mrs. Temes the camera. She grabbed it and like a true merchant, she bartered it for Zsuzsa's virtue. The Russian even smiled as he carried it away. Zsuzsa was safe.

Not that it matters, but it is pertinent to recall that Mrs. Temes did not make good her promise. In the months that followed we starved and were in a bad way. On the rare occasions that I saw her she never stopped to talk to me. I did not consider that she owed me anything and I just left it at that. It was a surprise, therefore, when almost three years later she called at the apartment the night before I left for England. She handed me a small, brown package. "It's for your trip" she said, then turned quickly and left. Inside the package was a chunk of *szalonna*, smoked pork fat, which, accompanied by crusty bread, formed all my meals for the next three days. I remember the customs man in London looking at it with curious disapproval but he let me keep Mrs. Temes's most useful gift.

The dead German continued to lie in the doorway. He was frozen solid now and we had no means of burying him in the hard ground. He was covered in spit and his uniform was beginning to tear at the hips and chest from the kicks administered by passing Russians. Once we were able to get out into the street we dragged him across to the railway lines, just so that we did not have to step over him.

When the Russians found out that my father spoke their language they took him away to interpret for them, so I saw little of him. He was away on the night of the 17th, perhaps just as well.

Strange new army road signs appeared in the Cyrillic alphabet and I amused myself by deciphering them. Since they gave the names of cities and towns I knew, reading them became easy.

Moving about was still risky. "Idyi suda!" became the first Russian words I learned. This "come here" could ring out any time, so we tried to give any Russian a wide berth. During the day this was easier, but given a determined soldier even crossing the street was no insurance. At night the words came out of the dark and the only way to avoid them was to stay off the street. Once you were thus cornered, out would come a gun and the next two new words in my vocabulary, "Davai chussi!" meaning "give me your watch." Once your watch was gone you had to turn out your pockets and it was "Davai" anything you carried that they fancied.

This taking of property by force, "Zabraty" in Russian, became one of the commonest words heard in those days and those who lived through the experience actually began using it with Hungarian suffixes. It is part of the language today.

Needless to say that when the front line is passing through there is no law. The law is the whim of the man with the gun. There is no appeal. Once the fighting has gone past, looting is not allowed, at least this is what we learned on registering complaints with Russian officers. However, nothing was ever done about complaints and there was no redress.

In fact, things got worse. Open looting began in broad daylight. As I stepped into the courtyard one morning I saw a Russian soldier and a woman, obviously a Hungarian, enter our apartment on the second floor. I hurried to tell my father. "Do nothing," he said. "If you resist now, they will come back later anyway. Just don't get hurt trying to protect what you cannot!" I watched until some time later they emerged, carrying large bundles made of our bed sheets. I immediately went to the apartment only to find they had taken everything of value.

Hunger now became the dominant factor of the day. It came in waves and at times was unbearable. Apathy would then take its place until the next wave would hit. I thought about food all the time. We still had only one meal of boiled dry beans or boiled dry peas a day. As the Germans were holding out there was still no food coming into town so we had to exist on supplies we had. The looting badly reduced our bartering power.

Worse was yet to come. Word got around that during the night the Russians laid siege to many houses in the neighbourhood and looted at will.

So far we had escaped this but we were advised to reinforce the front door to prevent forced entry. We got to work and soon had thick beams wedged into the wall at night. When they were in place the door could not be opened.

I realized that not being able to speak Russian was a great disadvantage to me, so I decided to teach myself the language. I managed to get hold of a "teach yourself" book and dictionary. The former was an excellent textbook of its kind and within six weeks I could read, write and speak essential Russian. At "Idyi suda!" I could tell them that I was a student and had nothing left to give them and would appreciate being allowed to go. The soldiers usually responded with good humour to a kid's attempt to speak their language.

On January 26 everyone went to bed early. The front door was reinforced and we did not feel threatened. The only movement in the night was Mr. Györgyfalvi's trip to the bathroom. He had a small apartment on the ground floor and used his own bathroom. I loved these trips. He was a short, thin man with a very large, hooked nose presiding over a lined face. He wore a floppy, pointed touque of a nightcap and a large nightgown to bed. On his sorties he would light his way with a candle set in a holder held by a ring. The approaching figure looked like a ghost, but I enjoyed the profile best of all as it followed the candle this way, then that. On this night, his trip over, he blew the candle out and I went to sleep. Voices woke me. Russian voices behind a flashlight. Everyone sat up and we were told to vacate the building by 8 a.m. There were no exceptions. The Germans were pushing back and the Russians needed our house for defence. It was about 3 a.m. We had nowhere to go and the deadline was close. We started packing what little we had with us, mostly bedding, some towels. The only clothing we had was what we wore. As first light came, my uncle set off to see an elderly spinster friend to see if she would let us stay with her while the emergency lasted. We were relieved when he returned saying she had agreed.

When we got ready to leave the Germans were shelling from the Citadel, in full view across the river. As I walked on the sidewalk with a load a shell hit the wall some ten feet above me, making a deep crater in the plaster. I was not hurt. Soon afterwards I was up in the apartment when a shell hit the wall between the two living room windows. Noise and dust, but no damage. I was lucky that day. We were off by eight, as ordered. We saw no one during our lengthy trip but the shelling did not come near us again. At long last we were welcomed into a small and oh-so-clean-and-tidy apartment by a portly, kind old lady, Piroska, which in Hungarian means "Little Red."

I hated the days with Piroska. For one thing, I felt we were not really welcome. For another, the flat had no view and there was nowhere to go, nothing to do. It was a prison without bars. Besides, meals were a pain. Piroska had a good supply of food for herself and clearly she could not share it with a bunch of newcomers. Our food supply was pitiful. Meals were an embarrassment on both sides so we ate at separate times and each party got busy with something while the other was eating. However, there was no discord.

One day we came by some flour and made dough for bread. A nearby baker let it be known that he would bake for anyone who brought in dough by 8 a.m. I set off with our dough, had it marked with a sticker and was told to be back at 10. When I returned, I feasted on the smell of freshly baked bread. I was so hungry I could have devoured the loaf on the spot. My conscience told me to take it home and await my turn. It was terribly hard. Hunger tempts one to the limit. I yielded, but only to the point of breaking off a one-inch piece of outer crust. I got home with my conscience satisfied but my stomach still empty.

Piroska had no fuel and we were cold. Our undernourished state made things worse. We had wood at our own apartment but that was in the war zone again. Still, we were going to try. My father and I set off with a couple of sacks. The Germans must have had a lot of ammunition parachuted in (we used to see the parachutes floating down at night) for the shelling was quite heavy. There was no one in the streets. As we approached our apartment building from the side street we heard heavy small arms fire in the main road. Just inside the side street stood a tall Mongolian soldier with a Maxim machine gun, sporting a very large drum magazine. He grinned from ear to ear. Whenever the shooting was particularly heavy he would fire off a burst in the air, laughing as he did so. Would he mind if we tried to get into our home round the corner? Mind? He roared with laughter and beckoned us with his hand to go ahead.

The shooting was a problem. Judging by the sound the main road was no-man's land. We might make it to the front door with no load but getting back to the corner with our sacks full was another matter. We had time, so we listened to the shooting, looking for some pattern which might give us a clue. Rightly or wrongly we thought there was a rhythm and a predictable lull, so we agreed to move on cue. It came quite soon and we ran. Twenty yards is a long way when you are in no-man's land. A fast moving figure in no-man's land is no man's friend and the whistling and the whining of the bullets around us served to press that lesson home. Back inside the front door things were quiet and the familiar surroundings made us feel good. We filled our sacks with firewood. The memory of the whistling bullets on the way in made the prospect of the return trip most uninviting.

Patterns in the shooting? We no longer believed in them. "I'll go first, I'm a kid, maybe they won't shoot at me," I said. My father agreed, knowing that the second person out is the easier target. The question: when? Make a mistake and it's all over, or worse, you may lie wounded where no one can help you. I found that easy. I thought only of getting through, so I got ready. How do you know when you should move? The answer is you don't *know*. You *feel* it. I felt it and I stepped out into no-man's land.

I saw no soldiers but knew that I was in the gun sights of both the Russians and the Germans. Maybe this was the nightmare of my childhood in which I was trying to run from an unseen pursuer, my steps hampered by deep mud. The sack felt like it weighed a ton. I could not run under its weight. I staggered with determined steps towards the corner, seemingly a long way off. A moment of truth, to be sure. Out of the line of fire, I threw the sack to the ground, waiting for my father. Strange, no shots. My father came running round the corner and we grinned and savoured what we had done. In truth, we could not have done it without the decency of the soldiers on both sides who obviously saw us go in, and, recognizing that we were civilians, held their fire on our return. Such gallantry goes unrecorded in the annals of war but as a survivor I tip my hat to both Russian and German soldiers who made it possible for us to have a hot meal that day.

The long haul back was not without problems. Shelling was still heavy. It did not bother us in the side streets. The main roads, however, radiate from the centre of town which put them straight into the line of fire of the big guns from the Citadel. I was well ahead of my father in the side street.

I then had to cross the main road and take the next left-hand turn leading to Piroska's flat. I crossed the main road. A few yards away, right in the middle of the sidewalk, there lay a wooden box. I sat down to rest on it, waiting for my father. He was still out of sight. Some ten or fifteen seconds later a shell screamed by and hit the wall about eight feet above ground. Jagged, red-hot splinters lay at my feet, sizzling in the cold air.

Had it not been for the improbably placed wooden box, I would have been just where the shell hit. Our judgment aside, so much in life is determined by chance.

My father, seeing the blast, dropped his sack and ran to find out if I was alright. I just sat on the wooden box grinning, so he knew I was.

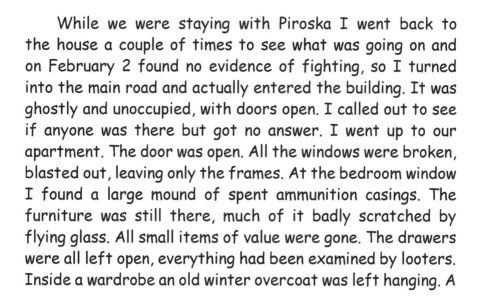

While we were staying with Piroska I went back to the house a couple of times to see what was going on and on February 2 found no evidence of fighting, so I turned into the main road and actually entered the building. It was ghostly and unoccupied, with doors open. I called out to see if anyone was there but got no answer. I went up to our apartment. The door was open. All the windows were broken, blasted out, leaving only the frames. At the bedroom window I found a large mound of spent ammunition casings. The furniture was still there, much of it badly scratched by flying glass. All small items of value were gone. The drawers were all left open, everything had been examined by looters. Inside a wardrobe an old winter overcoat was left hanging. A few thin towels remained. The eiderdowns were still there, along with a blanket or two. It was time to return to the house but we still had to stay in the shelter as the battle for the Citadel and Royal Castle was still raging.

We thanked Piroska for her kindness and returned to the shelter. The fighting could not last much longer, so we resigned ourselves to waiting. The prospect of living above ground soon bolstered our spirits. If only we had food!

One morning a man came saying he had flour to sell. My aunt traded her overcoat for it and I carried the sack down to the shelter. Food at last! My aunt got down to making dough for bread. In just a few hours we would have fresh bread! I gathered wood to fire the kitchen stove and found new motivation at the thought of eating again. By now I felt quite weak and found it hard to keep warm. Food would help all that.

I watched with anticipation. She had a deft hand at this and the dough was forming well. However, quite soon she found it increasingly sticky until it hardened altogether. The rascal who took her coat gave a sack of plaster of Paris with a thin layer of flour on top.

The city fell to the Russians on February 14 after a siege of 59 days. We left the shelter and moved upstairs. It was still very cold. We had no windows and the wind blew right through the apartment. We spent all day in the kitchen, the only room sheltered from the wind. I foraged for bits of wood all day to keep the stove fire going.

Being cold was now as bad as being hungry. I ran errands for people for a couple of spoonfuls of sugar. One of my jobs was to deliver a warm lunch to a man a mile away. For this I was paid about two ounces of granulated sugar. When I delivered the food, I could feel the heat inside through the open door and longed to be asked in if only for a moment but the door always closed quickly and I stayed in the cold outside, waiting for the container.

One of our neighbours had plenty of food, they even had chicken once or twice a week. The wife asked me one day if I would kill the chickens for her in exchange for the blood. I abhor killing anything, but I was so hungry that I readily agreed. I would do it sitting in a chair, the chicken between my thighs facing away from me, with my forefinger behind the animal's neck. I would then bend the neck backwards with my thumb and slit the poor bastard's throat with a sharp knife. The skill came in rotating our two entwined bodies quickly in a way that would direct all the spurting blood into a mug. I would then run home, fry the blood and eat it. I always shared it, so we had some protein. A gruesome way to get food but when you are starving the game changes.

Food was all I could think about all day. Waking up hungry and knowing there would be nothing to eat until six in the evening was tough. We still lived on boiled beans. On good days we boiled them with salt in the water. I made pointless trips back to the shelter just to go by the Temes' apartment and smell their food on the way.

Since we had no food I had to look for food. I reasoned that badly bombed, abandoned buildings must hide some food in collapsed kitchens or pantries. I therefore began to explore ruins in search of hidden treasure. These unoccupied buildings were partly or mostly collapsed and getting to the remnants of upper stories required some tricky climbing. The motivation for food made me forget all about the risks.

I had qualms about taking food which belonged to someone else, but actually such food as remained must now be four or five months old and whoever owned it had been dead or gone that long. I must say I never looked for anything other than food and even though I was badly malnourished, I never was tempted to loot or steal.

One morning I found treasure, three floors up in what had been a kitchen. The two outside walls were gone and looking down was a bit sickening. There was a slightly damaged cupboard against the inside wall. All its doors were open but one. I felt a surge of excitement as I opened it and there it was: food! Half a loaf of bread waiting for me. It was rock hard and covered in mildew and grit. I tried to take a big bite out of it but it was too tough. At length I bit off a piece and sat there chewing it endlessly, mildew, grit and all. It was a feast. The house had been bombed on September 5 and it was now mid-February. I took the loaf home with me. No one was interested in tasting it. My parents choked me off for risking injury in bombed buildings and they did not like my taking anything that was not ours. I ate the whole loaf myself and from then on I kept quiet about my expeditions.

With food getting more and more scarce I lost my job as a contract killer of chickens. No more commissions for running errands. Besides, I was getting pretty weak now. Wood was impossible to find, so we only fired up the stove to cook. The rest of the time we just sat, trying to keep warm.

I slept on the sofa in a room with no windows. The weather was still about -10C by day and colder at night. Going to bed was a trial. I had an eiderdown to cover me which was good. However, undressing in the cold and then waiting for the eiderdown to trap enough heat for comfort was hard to take. It took about an hour before I was warm enough to think of sleeping. Once warm, one had to stay in one position, otherwise the cold got in again.

Turning from one side to another required rotation about one's axis, lest the heat escaped. Getting out of bed in the morning and washing in cold water brought no relief.

I had a long walk one day as the weather became warmer. There was still some ice on the road but the bite had gone out of the cold. I felt as if I had been freed from prison after two months underground. A Russian approached on horseback. The horse had a mind of its own and the Russian had trouble keeping him under control. The horse wanted to run but the Russian kept him on close rein. The struggle between them was obvious as they approached. The horse reared and the Russian swore just as they got close. The rider stayed on. He unzipped his tunic and pulled out a brown bag. He looked inside, cursed and threw the bag into a snowbank. Horse and rider were gone in a canter. I waited until they were out of sight and then opened the paper bag. Inside were twenty-three eggs, three of which were broken. A windfall! I hurried home and we had a feast.

Good things could also work to our disadvantage. My cousin János visited me one day. We had a lot of experiences to compare and we wandered off into some deserted railway tracks along the river. I wore a pair of shoes that had been through the "knee deep in liquor" experience. By now they had lost the smell of liquor that I had found nauseating for so long. János had a pair of pretty good German army boots which he borrowed from a German whose head had turned into a pancake under the wheels of a Russian army truck. We were oblivious to the solitary Russian holding a rifle at us until he cried "Stoy!" (Stop!) He motioned to János to take off his German boots. János complied. The Russian lay down his rifle, took off his own boots, replaced them with the German boots and seemed pleased with the result. He picked up the rifle and left us without saying a word. János had to make do with the Russian's boots which were very uncomfortable and leaked besides.

As the weather warmed up I began to feel progressively weaker. On waking one morning my jaws were terribly painful, as if I had hot coals in my mouth. It took all my effort just to get dressed. I staggered into the kitchen and slumped into a chair. My mother took a look at my mouth and saw my gums bleeding. I felt so weak that they moved a sofa into the kitchen for me to lie on. No more expeditions or errands, this time I was down for the count. Our next door neighbour's daughter came and sat by me. She even held my hand and was awfully nice. I loved her for it but in my weakened condition I must have been pathetic company. My mother got me some food, from where only God knows, but my jaws were so sore I could not chew. I just wanted to lie down and die. I had no strength left to mount any sort of counterattack. Soon I could hardly raise my limbs, let alone my head. My mother told me I had scurvy but we had no access to vitamin C to reverse it. My father had a cousin who worked for a chemical firm and he set out to track him down. It took a week or so in the aftermath of the siege. By that time the neighbour's daughter was looking at me with great pity and I thought my time must be up. Without outside help I could go no further.

At last my father came home with a bottle of vitamin C tablets. I took a lot of them. They burned my mouth and my stomach, but within a couple of days I began to feel very much better and was soon up and about again. I wanted to thank the cousin but he, hearing of our dire circumstances, stayed away for good.

I took a long time to recover from the scurvy. My gums bled for more than a year. Cuts and bruises I sustained failed to heal for months. In the fall of 1945 there was a general election. Listening to all the platforms I threw in my lot with the Smallholders' party. I undertook to put up posters and stickers for them. Every afternoon during the election campaign I reported to party headquarters with two of my friends. We were given a bucket of glue and a brush, along with posters and pocketfuls of stickers with the party emblem, an ear of wheat, and we would set off on foot to do our work. As soon as we left the headquarters we were joined by goon squads whose job it was to taunt and threaten us, following us everywhere just a few yards behind. If we got on a tram they got on with us. If we suddenly got off, they did, too. When we were finished they split up and followed us home. On election day they waited for us as we got out of church, mounted the sidewalk with motorcycles and gave us a thorough beating.

With the election over, I had hundreds of unused small stickers left. I had cuts over each ankle that had failed to heal and continued to fester for months. I scrounged some gauze from God knows where, split the stickers lengthwise and used them as "band-aids" for weeks on end until my wounds stopped festering.

With the arrival of warmer weather the city came to life again. Along the circular boulevard there flourished a lively black market. There were "fortunes" to be made, so I was told, so I tried my luck selling cigarettes. Prices fluctuated and could vary considerably from one end of the market to the other, a distance of some twelve blocks. The secret was to arrive early, cover the whole stretch, buy at the cheap end and sell at the other end. Starting with very modest means I had to do a lot of running to make ends meet.

Running proved to be a more difficult problem than I had imagined. By this time my only shoes had disintegrated, so I took to wearing a worn pair of canvas gym shoes I found along the way. Very soon these had a hole in the sole, so I made plywood insoles to eliminate the defect. I was pleased with the resulting comfort. After a few days, however, the plywood inserts wore through the canvas and as I got into high gear running as a black marketeer, the insoles wore through the canvas tops and got left behind. Every 50 yards or so I had to stop, pick up the insole and shove it back through the slit in the heel and keep repeating the procedure to the point where I lost the advantage of speedy movement and, with it, my effectiveness as a black market cigarette dealer. In any case, I knew I had no talent for business, so it was just as well.

About this time inflation was in full swing. We began to talk of hundreds, then thousands, then millions and eventually trillions. The printing of new money could not keep pace with the escalating inflation. I realized that my gains were being forfeited so I looked to some investment of more lasting value. One of my classmates, Tibor, seemed to be very knowledgeable about the world. After all, it was from him that I first learned about the more lurid aspects of sex, so I consulted him. He had the answer, as I expected. He was in possession of a valuable stamp that would maintain its value, no matter what the rate of inflation. I bought the stamp and felt good about it. Later when I tried to sell it to a dealer I was told it was worthless.

Interestingly, a year later I ran in a one hundred metre race against Tibor. I began to come on at about seventy metres when he veered into my lane and stuck his elbow in front of my chest. I came second. The gym master told him he ran well. I was so disgusted I never ran another race. Complaining, I knew, would do no good. You just have to know how to take your licks.

Some time later Tibor got into mischief and the school took serious disciplinary action in the matter. Those of us who knew him well were formally questioned in private about his activities. I had a most difficult time under questioning since I was conversant with the issue. I knew that the school was right in going after him, yet I could not bring myself to reveal the facts that would condemn him. He had already failed me with the stamp. He failed me in the race. The conflict in my loyalties was hard to bear. Decades later it still is.

As the ground thawed with the arrival of spring, the thousands of people buried in shallow graves in city parks were exhumed for burial. It was there that I learned that there is no stench to equal the smell of rotting human flesh. Strangely, near the exhumation, someone was selling homemade pastry at a stall. For some months afterwards the thought of pastry brought to my mind the rotting bodies I had seen.

The arrival of spring 1945 was exciting. In a way it was like a rebirth. While the fighting and bombing were going on, the thought of dying was never far from one's thoughts. Now there were no more air raids, the sirens were silent, no shots were fired and the sun began to thaw us out.

The latter was the greatest gift of all. At last one did not have to shiver in inadequate clothes and going to bed was no longer an exercise in survival. Gone were the days when I delivered something on an errand, hoping they would invite me inside for maybe just a minute, only to have the door shut in my face, leaving me out in the cold. So at last, we were warm. One has to go without heat through a cold winter to appreciate the kindness of the sun.

We were warm, but we still had no food. Our supply of beans was running out. Toward the end we ran out of salt as well, so our single meal of the day was a plateful of boiled beans at six in the evening, very tasteless and very boring. When the beans finally ran out, we existed on a boiled corn mush -- it was no better than the beans except when we could improve the flavour with fried onions.

By this time my scurvy was better. My gums still bled and my cuts were not healing, but I was strong enough to go in search of stinging nettle to provide a source of vitamin C. I was never interested in plants before but now I could spot stinging nettle a long way off. When cooked, it tasted like spinach.

Of course, with the cessation of fighting all excitement died down as well. There was no radio, no library, nothing to read or to listen to and, for a fourteen-year-old, sitting with the family all night was pretty boring.

It was a blessing when school finally opened again, although only for a couple of months before the summer recess. I had my worst grades ever but somehow I just could not get interested in learning. I was still busy surviving. My friends appeared well fed and clothed. In contrast, what the bombs did not destroy of our belongings, the looters took away. We were left with what we were wearing. I felt like an outcast among my friends.

During the short school period the Danish Red Cross set up a daily food kitchen at our school and we were treated to a plateful of porridge, with sugar, each day after school. Thank you, Danes -- it was delicious!

Now that the fighting was over my friends got me interested in exploring the Citadel. This fortification overlooking Budapest from Gellert Hill, was one of the last German holdouts in the battle for the city. It was taken after a long, fierce battle, some of which I had watched from our living room window. Rumour had it that there were underground tunnels still full of weapons and dead Germans. I could not pass up a trip like that, so off we went. The place was home ground to me as I had hundreds of walks on the hill in my childhood, though I had never been inside the Citadel. The sun was high and a fierce wind was blowing, whistling and whining as it eddied among the ruins and rubble.

The Citadel is a high, semi--circular fort, built of white stone with a large inner courtyard. It dominates the city across the river and from it the view to the North, East and South is limitless. To the West there are more hills.

The courtyard was deserted. From it there opened dark passages with no doors. These must be the tunnels we had heard about. After some hesitation we entered through a dark archway. Daylight stayed with us for only a few yards, but once our eyes became accustomed to the darkness we were able to proceed for quite a way. From then on, we had to feel our way along the wall, turning back frequently to identify things against the receding pinpoint of light that was our entrance. I had expected to stumble on dead soldiers and machine pistols but there weren't any. It was totally dark now where we were and we could go no further without a source of light. Just then, looking back toward the entrance we saw two figures enter the tunnel. We thought they were Russian soldiers but could not be sure. What next? We had to make a quick decision as they were approaching fast. It would be a disadvantage to meet them in complete darkness, they may shoot when surprised, so we decided to move quickly back toward the entrance, talking loudly, so as to forewarn them. They stopped and waited for us to emerge into the half light. One of them held a long-barreled revolver.

By now I had been held up by Russians so many times that I was not unduly worried, though this could hardly be said of my friend. "Zdrastvuytye, tovarishchi!" I called out in cheerful greeting. "Zdrastvuytye," they replied and then continued, "Davay chussi." As it happened, my watch had been taken long ago, while my friend's watch was on full display. He reluctantly parted with his watch while the gun was turned on me. "Chussi," said the Russian quietly. I remained upbeat and cheerful and explained to him in my self-taught but quite adequate Russian that I had no watch because I had already "given" it to a Russian soldier. I was careful to avoid the verb "zabraty," implying robbery at gunpoint, as, with the gun barrel pointing straight at me, I thought it was a time for discretion. He nodded and used the gun barrel to indicate that he wanted me to turn out my pockets. I had a dirty handkerchief in one and my fountain pen in the other. He took the pen, then motioned with the gun for us to walk ahead of them out of the tunnel, back into the courtyard. It was an uneasy walk, not knowing what would happen next.

Squinting in the bright light again, we were met by howling wind in the courtyard. We could not hear the Russians behind us, and, in any case it was time to stop and find out what they wanted. I turned, only to see them running away. By then the Russians were an occupying power, not a fighting force, and they were not allowed to interfere with and rob civilians. This I knew. I nudged my friend's elbow and took off in pursuit of the soldiers. As I got near, the one with the gun heard me. He stopped, turned and took aim. One learns discretion very quickly under such circumstances. I shrugged my shoulders and turned away. The soldiers were soon out of sight.

My friend was fed up with the loss of his watch. Time to put my fledgling Russian to use. I told him we were not going to leave the matter at that. I suspected that these soldiers were from the anti-aircraft battery stationed halfway down the hill, so we headed down there to speak to the sentry.

"Zdrastvuytye, tovarish" I said, using the well-worn formula. "Zdrastvuytye," replied the sentry, amused that I should address him in Russian. The soldiers always responded to kids with good humour.

"I want to speak to an officer," I said. "An officer?" he said. "Whatever for?"

"Two soldiers robbed us up there," I replied, pointing at the fortress, not knowing how to describe it in Russian.

"Ah!" he said, shaking his head. "Wait here!" He turned and walked off in the direction of the big tent.

My friend was deeply impressed. I felt very smug myself, recognizing my new role as a fluent "bilingual" advocate, maybe even a candidate for the diplomatic corps. We waited in the wind. The longer we waited the more convinced I was that my approach would bear fruit. My friend even stopped fretting about his watch, knowing it would be returned to him in a few minutes with a small package of the ground-up remains of our assailants and a written apology from the commander of the Russian garrison. Knowing we were in the right gave our confidence wings.

At length the tent flap was opened and out came a Russian, then another, and another, until there were at least twenty of them. They all carried weapons. Surely they could not all be sentries! The war was practically over!

They stopped some twenty yards from us, cocked their weapons and began firing over our heads, all at once. They kept it up. This was not a day for diplomacy. Might was right, and we knew it. We turned and did not run, knowing they were just making a point firing over our heads. All the same, we walked away rather quickly.

CONCLUSION

The guns fell silent. The armies marched on. We emerged from the cellar to witness a scene of devastation we could not have imagined. Visited by the apocalypse, the city lay in ruin. Maybe this was the end of the world. The once proud bridges across the Danube lay twisted and contorted in the water. Buildings still standing bore scars with windows gouged out, roofs hanging, walls gone. Shell craters and bullet holes punctuated the view.

Rubble rose high in the streets. The walls of burned houses stood like skeletons in this cemetery of doom. The air was heavy with the smell of fires long burned out. There was no fresh air to breathe.

Yet, all was not lost. The fear of the unknown had finally left us. The wild interplay of life and chance, despair and hope, had ended at last. Of necessity, we had to make a fresh start. Sure, we endured hardships but we regarded them as tests passed. There was no point in looking for redress. We had lost everything of value and emerged with only the clothes we wore. Still, we retained the most precious possession of all: Life. We had to remember that the cold, the fear, the hunger, the de-humanizing lack of sanitation, the buckets, the lice we endured were still signs of life and, through them, we paid our dues for survival.

Danger survived becomes experience. It gives renewed perspective on challenges that visit us later. Seemingly big problems become trivial in comparison. No threat is quite as intimidating the next time. The knowledge that you can emerge unscathed from hopeless situations becomes a source of optimism, of positive outlook.

These thoughts gave me comfort, even though our hardships were not over. Of the basic necessities of food, warmth and love, the first two eluded us for at least another year. We endured another winter without windows. We did not mourn the loss of our possessions - in the turmoil we were bound to lose them anyway. We drew solace from the thought that we did not survive at anyone's expense. We learned to appreciate small comforts, a warm cup of tea held in the hand, a few minutes spent sitting in the sun, a hug from a family member that told us love was not extinguished in the storm.

The isolation we experienced as death began to stalk the streets receded as imperceptibly as it had come. It served a useful purpose for me as it gave me a chance to take stock of myself. I realized I could function on my own, follow my instincts and acquiesce in decisions I made because in the circumstances they were the best I could reach. I told myself that my poverty was only a starting point for better things later.

Life is like a journey on a train: whatever class you occupy, you end up at the same destination.

I think of Nikolai now and then. I harbour no hatred for him. Like the rest of us, he was part of the flotsam of a great storm. He, in his own way, probably suffered as much as we did.

At the outset I said these were events I would rather forget. Now that I have committed them to paper, maybe I will.

ALTAR EGO

In Roman times Christians were thrown to the lions for the entertainment of the crowds. Early in my career as an altar boy I learned to appreciate their predicament.

By the age of eight I served Mass regularly at the convent school for girls. Although I did not understand a word, I knew by heart the responses in Latin. This gaveme a degree of portability and even periods of high demand when there happened to be a shortage of altar boys.

Yet, it was not all plain sailing. Some skills, essential to my progress, remained to be learned. The most immediate challenge was the art of serving Mass on the right hand side of the altar. The server on the left did the menial jobs, like carrying the book from one side of the altar to the other and pouring only water, not wine, over the fingers of the priest. The priests themselves usually inclined to the right to catch the responses of the right hand altar boy, ignoring my equally fervent whispers from the left.

One, only one, priest ever acknowledged my presence while I was left hand altar boy. Endowed with a wedge-shaped head and uncommonly acute hearing, he alone discerned the fractures and imperfections of my fledgling Latin and would impart his displeasure in subtle ways, totally unobserved by the faithful in the nave: "GRATIAS AGIMUS TIBI….you'd better learn these responses soon, my son, this is nonsense!……PROPTER MAGNAM GLORIAM TUAM."

George was the senior altar boy, the favourite of the nuns. He was the child they never had, with curly red hair parted at the side, a long straight nose and a perpetual smile. What's more, he did spend a moment or two in devotion during the quiet moments of the Mass. At such times the nuns must have seen him as a cherub come alive from the frescoes above.

I set to work. I polished up my Latin responses. I began to build a view of what was happening. Without such understanding one could not ring the bells. Yet, if ever there was a glamour job at the altar it was not the Latin phrases, not the pouring of the wine, but the ringing of the bells. Oh, how I waited to get my hands on those bells!

Practice was out of the question. The nuns would not have it. Not in the sacristy, not in the church. I just had to watch George a little longer.

Nor was the physical task of ringing the bells the only hurdle. The timing had to be right. I could not learn it from a book. The clues came from the gestures and genuflections of the priest. I watched the priest as much as I watched George. The pieces of the jigsaw fell into place. Soon I would get my hand on the bells! At last there was no doubt: I was ready. I must talk to him.

I found him in the sacristy, busy preparing the vestments for High Mass. He moved quietly, with the deliberation of one who is on top of his job. He acknowledged my greeting with barely a nod and went on working. The church across the door was still in darkness save for the light of a candle at the altar. An unseen organist played in muted tones. There was no one about. I entered the church and from the safety of a pew in the back I took stock of the situation. The gate of the altar enclosure was closed. Around the enclosure the chairs for the senior girls had already been set up. The candle over the tabernacle flickered reassuringly. The quiet organ music made me feel relaxed. And there, on the extreme right of the second step, were the bells. I felt my grip tighten instinctively. I rose and headed back to the sacristy.

With all the pent-up resolve known only to those who are shy I strode up to George and without waiting for him to notice me I informed him that I...would...like...to serve on the right...side this morning. He put the finishing touches on his job and, still without acknowledging me in any way, stood back to give it final thought. Just as I scraped my throat to ask him again, he turned from the vestments, broke into a smile and informed me that it was a marvellous idea.

I was elated. For a minute or two I could do little as my sudden joy left no room for thought. I wanted the world to hear but there was no one about. Even the unseen organist had melted away - the church was now silent. In the corner of the sacristy a clock was ticking. George stood aside, detached, smiling.

At last he broke the silence. His voice was businesslike, his phrases pert. Yet, his manner was different. All at once, he talked to me on the level - no longer addressing the left hand altar boy but rather an equal. I did what I could to hide my pleasure but I was very flattered.

He went on to explain that it was a fortunate coincidence that I should assume the role of right hand altar boy that day, since it was an anniversary of sorts in the life of the convent. Many people previously associated with the convent were returning for the occasion.

He briefed me on certain points. The priest liked a generous quantity of wine poured from the little jug . Going through the church to and from the altar we were to walk slowly and in step. We could not, under any circumstances, acknowledge anyone. The bells must be rung slowly to bring out their resonance. This was best judged by actually listening to the bells. Better yet, in case I had trouble George had a foolproof safety measure.

My spirits soared. Inwardly I still had a gnawing insecurity about the bells. Whispering the Latin responses and pouring the wine were discreet skills veiled by silence, but the altar boy declared himself when he rang the bells. The people stood, genuflected, even beat their breasts to their signal. The harmony gave extra dimension to their devotions. The bells must be rung properly and George was going to help me do it.

His method was simple. He would kneel with his palms together in the attitude of prayer. When it was time to reach for the bells, his hands, still held together in prayer, would point downwards. Each time I had to ring the bells the hands would point upwards. That was all. Simple. He really was the Master.

The nun in charge of the sacristy arrived and gave George a serene smile. She gave me no greeting. With her arrival all conversation ceased. She prepared the chalice with its embroidered coverings and lit the candle on the table bearing the priest's vestments. Her full, long black habit rustled purposefully as she moved about. Her rosary chattered quietly in accompaniment.

By the time I lit the altar candles the church was almost full. The senior girls were filing in. The organist was playing again.

I turned to look at the scene from the sacristy door. A tremendous setting for my big day. Before those candles were extinguished again I would become a fully-fledged altar boy. Nothing could hold me back now, nothing!

Whispers in the sacristy told me the priest had arrived. I backed in quietly and closed the door. At that precise moment lightning hit me - or so it seemed. A horrible emptiness seized my stomach and the sparkling image of my elation of a moment ago lay in shattered pieces on the floor. The priest had, indeed, arrived. Of all the priests it had to be the one with the wedge-shaped head and the uncommonly acute hearing.

I quickly signalled to George that the whole thing was off, that it was all a big mistake, premature at that. He just smiled and shook his head. The priest was now in prayer and communications were cut off. I managed a final "but George....." when George signalled that the priest was ready. Then I knew it was too late.

The nun held open the door and we entered the church walking slowly, in step, with me on the right. The priest followed closely, murmuring his prayer. The tempo of the organ quickened, as did my pulse. The beautifully lit altar enclosure now appeared as the arena where I would face the lions.

I need not have worried. The initial exchange in Latin went well and the priest actually inclined to the right in what was obvious approval of my new status. George smiled in approval, the candles burned with intensity, the organ played at full volume and I was as happy as a right hand altar boy could be. There, before me at last, were the bells!

With the sermon over, we were fast approaching the point when I had to ring the bells. I glanced across at George,

but he was not signalling. His eyes were closed and he was obviously in the middle of his devotion. I tried a half-hearted "Psssst………" which got me nowhere, although a couple of senior girls did look up. I glued my eyes on the priest. No doubt this time. I had to ring the bells. George appeared to be sleeping. I slipped my hand into the grip, inclined the bells forward and with a slight swing of my forearm rang the bells once. My hand, my arm, my entire being resonated with the sound. It even woke George. I had broken the ice and had done so without help!

I poured the wine with the generous hand of one who is practised at the job. The priest's hand did not linger for more. Soon the bells must be rung again. Of course I could do it! What's more, I did not go to sleep on the job. I cast a casual glance at George, as if to tell him so.

The service was now entering its most mystical phase, the central events in the liturgy of the Mass. Although I did not understand them, I recognized their significance by the devout response of the congregation and by the unmistakable treatment given by the organist. What's more, I was keenly aware of the role played by the bells in signalling these moments.

"Sanctus, Sanctus, Sanctus" I suddenly heard the priest say, each word louder than the one before and with a peculiar, pressing emphasis on the second syllable. I shot a glance at George's hand for the signal. There was none.

"Sanctus, Sanctus, Sanctus," repeated the priest with a tone of mounting urgency. I grabbed the bells and rang three times. The priest went on with the Mass in whispers. I looked at George. His hands pointed down. That was my signal to get ready. I hung on to the bells for all I was worth. George looked at me with disapproval. His hands moved up and down. Thinking I was late again I rang the bells three more times. George looked upset. He shook his head. He motioned downwards. I rang the bells once again before I realized that he wanted me to put them down.

The organ stopped playing. The priest bent down to kiss the altar and as he did so he shot me a devastating glance under his armpit. The sound of disorganized movement came from the congregation behind me. Some knelt down, others obviously viewed my effort as a false alarm, and remained in their seats. As I could not see them, they did not bother me. Much, much worse for my self-respect was the reaction of the senior girls who were in my view only a few feet away. Some of them were maintaining a semblance of composure but one or two were sobbing with laughter they had tried to suppress but could not.

Worse was to come. The central, most sacred part of the Mass was upon us and I had to ring the bells altogether ten times. I wished that George would come to my rescue and take over but instead he was now signalling in earnest. To my immense disadvantage, I could no longer remember which way the signal went. I had lost my bearings. The organ was still silent. People behind me were now shuffling audibly in the silence and the laughter of the two senior girls to my right stirred my confusion.

I picked up the bells and started ringing. If George moved his hands up, I rang; if he moved them down, I rang again. There was audible adverse reaction from all sides and more senior girls joined in the laughter. Still I went on ringing. The priest turned to look at me for a brief moment, then went on with the ritual, sticking to his job under fire.

The bells grew heavy in my hand but I could not release my grip. George, though concerned, was still composed. His signals were getting through clearly but he did not realize that I had lost their meaning. The ringing went on without any relation to what was happening at the altar. Even the organist started up again as if to suppress the untimely resonances issuing from my hand.

My defeat was total. The Mass had now reached a stage where the bells were no longer required. I let go of them but felt no relief. The reckoning, I knew, was yet to come. I wanted to vanish without trace but the only exit was down the aisle. I had to face everyone during the long walk back to the sacristy. Once there, I had to face the priest.

The rest of the Mass and the slow walk along the aisle are just a blur in my mind now. I got out of my white robe as quickly as I could, vowing never to wear it again. My sense of duty kept me waiting for the priest to deliver his judgement. He ended his prayer, half-turned, addressing no one in particular and said: "A while ago the barber let his apprentice cut my hair. I made clear to the barber that it must not happen again!" A slight nod to the nun and he was gone.

Many years later, as I walked by the Colosseum in Rome, it was not the Christians and the lions I thought about. Rather, I remembered that distant Sunday morning on the right hand side of the altar.

NIGHTWATCH

My friend Guszti had a stop-watch, the object of admiration, of longing and sometimes envy for us boys who spent part of the summer of 1943 in our school's summer resort on Lake Balaton. The grounds were large, the vegetation lush and the nearness of the lake saw to it that we had refuge from the summer heat. The food was delicious, games of all kinds provided and we even had a fine sports field. The boys were accommodated in a dormitory in the two-story building where, with discreet segregation, our teachers also stayed. They had the luxury of a high terrace with a view of the lake where in the evening, after the boys had been sent to bed, they enjoyed the wines and cigars that were their just rewards. At least we were so informed by our keen sense of smell. Their conversation was beyond our hearing.

Guszti arrived with his own, pyramidal two-man tent and was allowed to sleep in it, not far from the building. As his alternate goal-keeper on the school's soccer team, I was invited to share the tent with him. It was akin to being upgraded to business class travel in today's terms.

He hung his stop-watch on the tent pole. No passer-by could resist gazing at the watch with its many hands and sporty design. In no time it became the central timepiece, a sort of "Big Ben", by which the boys navigated through the carefree hours of each day. One thing I particularly liked about Guszti was that as soon as his head hit the pillow he fell into deep sleep. The flaps of our tent were always left open and, lying on my back, I could listen to the song of a myriad crickets and watch the bright stars above while on the ground a thousand glow-worms lit the way.

For a city boy, whose parents could not afford a holiday, this was a time to cherish, an experience to remember. I tried to stay awake as long as I could, inhale the scent of flowers and the smell of the earth, making them part of me, this other me who did not want this idyllic time to end.

The city and its bustle, its clamour and pace was not a credible alternative to this untroubled existence where the sun held rule and the moods of Nature, the wind, the rain, the storms paid me personal visits at their whim.

If only time would stand still and I would awaken to find that our confined city apartment, our straitened circumstances and the constant reminder that the things we wanted were beyond our means were just a bad dream, soon dispelled by the breezes that blew in from the lake, tussling my hair!

An even darker shadow was fast approaching, a storm beyond my comprehension, still distant but its advance evident from every conversation, from the mandatory blackout at night, the proclamation of martial law and by posters on walls declaring death penalty for crime committed taking advantage of the imposed darkness. Perhaps this was the last joyful summer we were ever to know. But being young boys, our attention stayed within the perimeter of the present. We were concerned with what was, not what might be. Yet, the tidal wave of war was on its way.

Guszti was asleep but I was still awake. There was no moon and the night was unusually silent. The crickets had the night off and no breeze stirred. The boys in the main building were quiet and there was no laughter from the terrace to tell that our teachers were even there. It was a time when you could listen to the silence and hear it as if it were a crescendo of sound. I listened to it and relished its gift. The night was at a standstill. Maybe the world had stopped turning.

At first I dismissed the brushing noise I heard, it had no business to encroach on the stillness. It did not belong, it could not be real, the world was asleep. I gazed at the roof of the tent above me and rejoiced at being able to think undisturbed.

Tonight I had the world to myself. It would remain so till sunrise but I'd be asleep long before then. I felt the weight of my head sink into my pillow.

And then I heard the brushing noise again. It lasted a second, maybe two. It came from close by. Some heavy object, close to the ground, I reasoned. Maybe some animal on the loose but I could think of no animal heavy enough to cause a sound like that.

The brushing sound returned. Whatever caused it, it wanted to remain concealed. I did not want to frighten it away, so I moved my head slowly. With eyes accustomed to the darkness by then, I peered toward the entrance of the tent and saw a crouching human figure. It was within arm's reach and too large to have been one of the boys. Whoever it was, he was not there with good intent. I knew that Guszti hung a flashlight on the tent pole beneath his watch. Slowly I pulled my arm out from beneath the blanket and reached for the flashlight. The intruder heard the movement, withdrew on his knees and rose to run away. At that moment I turned on the light just in time to catch a glimpse of a well worn, shiny jacket with stripes. The figure vanished into the darkness. I jumped out of bed, shook Guszti and told him to follow me.

"What is going on?", Guszti asked, put out that his sleep had been disturbed.

"Just follow, I'll tell you later".

"Come on", he said, "what's all the fuss?"

"Someone tried to steal your watch!"

The latter statement galvanized Guszti into action and he was now wide awake.

Anyone wanting to leave the resort would have to pass by the sports field to the gate. As it happened, war regulations stipulated that every piece of land had to be planted with food crops and the sports field was covered with a tall stand of corn. We headed towards it, flashlight leading the way.

By now I knew we were dealing with crime attempted during the blackout. The intruder was also aware of the penalty, so he may become violent. I picked up a rugged rock and gave it to Guszti to use as a weapon in case we needed it. I chose one for myself also.

We could not see anyone, nor heard sound of movement. We stood by the side of the field and I turned off the flashlight.

"We'll wait, in case we hear something", I said quietly. Guszti, actually interested now, agreed. The darkness around us was total, still embracing the silence I so enjoyed only minutes before. But this time we were on a hunt and felt the excitement of a new adventure. We stayed quiet, as did our intruder, if indeed he was still nearby.

"I was mistaken, it must have been a dog! Let's forget it and go back to sleep", I shouted to Guszti, although he was standing right beside me. I hoped that my loud remark would carry and make our quarry move. We pretended to walk back toward the tent but listened for sound.

Sure enough, we heard the rustle of someone walking through the field of corn. Flashlight blazing, we ran toward the sound. In no time we came face to face with a man whom we recognized as a daytime worker at the resort. I told him we knew who he was and that he had better show up for work in the morning. He turned and left without a word, knowing he was a candidate for the rope.

We walked up to our Head Master's quarters and reported the incident. The man turned up the next morning and had an earful from the Head Master, who was not a lover of the war, much less that of capital punishment. The man was fired. He walked away with his life but without Guszti's stopwatch.

The watch continued to hang on the tent pole, as before.

IN GOES A BOY – OUT COMES A MAN

"Men fear Death as children fear to go in the dark" wrote Francis Bacon. When I was a child I had such a protected existence that I never met the dark face to face. It held no dread for me.

All this changed during the war, when, to escape the bombing, we moved out of town to a village reached by local railway. I was fourteen. By late September it was dark by the time I got off the train on my way home from school.

Leaving the station I could take a lengthy roundabout path skirting the village or opt for a shortcut through the cemetery adjoining the station. At the end of a long day the cemetery was a more practical choice.

The villagers must have been well-to-do people as just about every grave was marked by elaborate masonry, marble heads, urns and life- size, shall I say human-size, marble angels. There was no formal path across the graveyard, so one had to be familiar with the tombstones along the shortcut and dodge the graves until the far side was reached. A piece of cake in daytime.

The cemetery was heavily wooded. Large willows predominated with a cohort of other deciduous trees, the leaves of which rustled constantly. Whenever the wind got up, the willows moaned in its passage while the rustle from the rest became a roar.

One night, tired and not a little hungry, I got off the train and headed for the cemetery gate. I stepped out with deliberate stride, knowing I would soon be home. The wind was strong, dominating all sound. It was pitch dark. I had the path memorized but once I got well inside the gate, I lost all visual reference. A blackout was in force. There was no moon, no stars. The next train was not due for a couple of hours, so there was no prospect of meeting someone who might show the way. I was lost. The chatter of the leaves at the mercy of the wind was enough to set my heart racing.

Even worse, I had to confess that I was well and truly afraid. At long last I came face to face with the dark and I was not doing well. Frustration rapidly evolved into shame. Not only was I afraid in these godforsaken surroundings but I did not know which way to turn. I could no longer tell from which direction I had come.

Worse was to follow. As my eyes became accustomed to the dark, hitherto unseen shapes drifted into view. The marble busts, urns and angels all assumed menacing human forms, every one a fiend barring my way.

The direction I randomly elected to follow led into the deepest, darkest part of the graveyard.

To make my plight worse, unpleasant recollections crowded into my head. Was it not just about here, only the week before, that a man I had known shot himself? Suicides always held a horror for me. That suicide took place in a dark recess of the cemetery. My location answered that description. Maybe one of those shapes I took for human form was another candidate for suicide within my hearing, if not within reach.

No one could help me. I had to extricate myself from this predicament. By now I was truly anxious. Yet, curiously, I did not panic. I felt the tombstones for guidance. They were disdainfully silent.

If anyone had asked me exactly what I was afraid of, I could not have given a plausible answer. Why should shadows resembling human form hold such terror for me? Was it because subconsciously I realised that evil lurked only in the hearts of men? The war was beginning to make that clear. Was it because anyone abroad during the blackout might be there with bad intent? Or was it simply that the things we fear most are those we cannot see?

Clearly, no one was following me and I knew I was going to be alright but the adventure had gone on long enough. I just wanted out!

More by luck than good judgement I found my way out of the cemetery and soon arrived home. I was too ashamed of myself to mention my experience and when asked at supper why I was so quiet, I just put it down to being tired.

All was not well. The humiliation I had suffered was more than I could bear. Something had to be done.

The problem was that I could not confide in anyone at home because I knew that they would just tell me to take the long way round from now on, instead of the short cut through the cemetery. As for my friends at school, talking to them about it would brand me a sissy. Those who would condemn me would not even have to prove that they could do better in the cemetery than I did.

As an only child, I always had to work things out for myself. I was my own counsel, my own judge, my own cheering section. When I was wounded, I tended my own wounds. You go through a lot of anguish before you acquire self-reliance that way.

I spent the next day pondering what I could do to restore my courage and, with it, my self-respect. I took the long way around in the evening, feeling not at all proud of myself. No one could help me, that much I knew.

After much heart-searching I concluded that the only way around my problem was to go back to the cemetery every night and walk around its darkest corners for a half-hour or so. I had to face the music. I saw no other solution. I realized that I would walk with fear at first but knew that if I persevered long enough I had a good chance of overcoming it.

Of course I could not tell the family why I was leaving the house every night so I just said I was going for a walk. The very first time I went, I put my things into order before I left the house in case anything should happen.

Entering the dark cemetery again I felt my apprehension return. Fear was tugging at my heart and my pulse seemed to race. Unpleasant business for sure. Still, I had a good feeling also, for I knew that if only I was equal to the task I set for myself I would lay this ghost for good. Situations are structured by our own thoughts and often magnified out of all proportion. I must make myself bigger than the situation, I reasoned.

My first return to the cemetery turned out to be uneventful. There was not much wind and the trees were not nearly as menacing as they had been. The headstones still gave me the willies but I was there to do a job and on emerging from the graveyard I thought I had done better than I had expected. I felt encouraged to continue.

The second night was much worse. I had actually started out quite well and was feeling smug about my new-found courage when all of a sudden I heard twigs cracking right behind me and as I turned I saw a black shape bearing down on me.

"Good evening," said the black shape and was quickly gone, twigs cracking underfoot.

"Good……evening!" I replied hesitantly, still catching my breath.

Subsequent walks in the dark cemetery became easier, just as I had expected. Now I felt tempted to explore the darkest corners. I sought them out and found them actually peaceful. No one attacked me and I found no threat in the headstones, busts and marble angels. I still felt though, that in order to graduate I would have to go back on a windy night. That would be the real test.

A howling wind soon arrived and found me traipsing among the tombstones. Strange, what had started out as an exercise to rid me of fear of the dark had, by now, become just physical exercise. A walk, nothing more.

I knew I had won. I still went back a few times after that but eventually stopped for an unexpected reason: it was boring.

This experience left me with a legacy of respect for the instinctive good judgement of young people.

After all, acting on the advice of a fourteen-year old, I entered the cemetery as a boy and came out a man.

ONE LAST LOOK

The summer of 1944 brought us face to face with the prospect of imminent death. The beautiful, sunny weather gave a clear view of the targets for the US Air Force by day and for the bombers, usually Russian, at night. Whenever bomber formations approached the country the radio went silent, save for terse announcements declaring alerts for the regions ahead of the path of the bomber forces, to be followed by full alarms which triggered the sirens, the foreboding sound of which became more vexing as time went on and we realized that death may be moments away.

Wisely, my family moved to a village, some twenty miles out of town to get out of harm's way. The house in which we lived had no electricity. Not having a radio, I built a crystal radio set, which, although the volume was faint, kept me informed of the daily threat from the air. There was not much to do in the village, so I amused myself by operating the village air-raid siren which, brought to life by a rotary handle, created enough trepidation to wake the dead in the neighbouring cemetery. Had it not been for the air-raids, there would have been little to hold my interest.

When the bombers arrived, anti-aircraft guns surrounding the city created a cacophony of sound, as disturbing for those on the ground as for the airmen above. The airplanes threw down bundles of tinfoil to put off ground-based radar and these shone with brilliance while in the air and resembled Christmas decorations when they reached the ground. Despite these measures, some bombers were hit, producing a disturbing sight as wings in flames broke off, cartwheeling in the air while the main body of the aircraft, spinning wildly, met the ground in a fiery explosion. I can still see a Flying Fortress in its steep, final dive with five bodies leaving it. Two of the parachutes were on fire. The bodies, accelerating out of sight, fell to ground. A sickening sight, even for those on the ground. Minutes after the bombers left there began a symphony of high-pitched whines, a thousand woodwinds playing high notes, as fragments from anti- aircraft shells fell from the sky.

Soon the village had its first victim: a young man in a distant town died in an air raid. With rail traffic disrupted by military priorities and the constant air alerts, it took several days before the body finally reached the village. Everyone turned out for the funeral.

As a boy of thirteen I managed to find a spot standing by the graveside between two tall mourners. The Protestant pastor did his job, while women in black sobbed and men stood by wearing facial expressions befitting the occasion.

At length the ritual was concluded and the time came to lower the victim into the ground to his final resting place. Four young men stepped forward, lifting the coffin with ropes while the wooden beams across the grave were moved away. Sobs and outright cries and wails intensified. Slowly the coffin began its descent. Some mourners even moved closer to steal one last look at the dearly departed son, husband, friend.

And then it happened: for whatever reason, one man let go of the rope, the head of the coffin rose at an angle, the lid flew off and the corpse, looking ghastly and very dead, looked back at the mourners. At that very moment the stench of decomposition exploded like a bomb on the hapless congregation, making everyone gasp in vain for fresh air. The crowd was driven back from the grave, as if by a blast and sounds of distress, cries, coughing and choking filled the air.

The pastor, to his credit, stood his ground and remained at the head end of the grave. A brave young man leaped into the hole and disregarding the horror of the moment, righted the coffin, closed the lid and emerged quickly, signalling to men with shovels to fill the grave with earth which they did with a speed not commonly seen at funerals.

The village returned to its routine, I continued to operate the siren for a while but ever since that day I have viewed funeral invitations with a modicum of distress.

EYE ON THE BALL

An impromptu soccer game in 1946 had me playing on the right wing. Play flowed back and forth. There were no stakes involved, not even pride. Just a bunch of teenage boys letting off steam. We had an old soccer ball to kick around while wearing our street clothes and shoes. The latter always bothered my conscience because those were the only shoes I owned. After a few games they were badly scuffed and less than presentable. There was no money for new shoes. But that was a decision for the head. We played the game from the heart.

We had no referee, did not need one. There were no fouls in a game between friends and we always seemed to agree on obvious off-sides. The ground on which we played had no lines. If someone touched the ball with his hand in the imaginary penalty area, no-one argued the matter. In fact, the score was of no consequence. What we lacked in skills we made up in spirit, just playing for fun.

This time we even had spectators. Two men in leather coats walked up and down the touchline, taking an obvious interest in what we were doing. Talent scouts? Hardly, there was no talent on the field. Still, they stayed and paid attention.

With play at the far end of the field one of the men called out to me. Very friendly, he asked me to go over to the touch line.

"Wish I could play!", he said smiling.

"Why not?", I asked.

"I'm past all that", he replied, as if in regret.

He then turned serious.

"Something bad has happened to your friend"

"What friend?", I asked, concerned

"See for yourself......" He reached into his pocket and showed me a photograph of a burned corpse. The body was unrecognizable. I shrugged my shoulders and shook my head, turning to see where the action was on the field. It was still away from me.

"It is not anyone I know"

"Take a better look", he said, this time showing a close-up of a face burned beyond recognition. I again shook my head.

"Here, you'll recognize him in this one", another close-up from a different angle. "He's your friend, that's why we are here".

"I tell you, I have never seen this man"

"Take another look, you know him!", persuading me to weaken.

In truth, the features of the burned face were not those of anyone I knew. Why was this man pressing me?

..........................and then the penny dropped......of course!...... leather coats!......Why did it take me so long?!

During the war the Gestapo agents wore coats like that and now that the war was over their successors in the Communist secret police did the same. They were trying to trap me into saying that maybe I recognized the dead man as someone I knew. If I weakened they would take me away as their star witness. I asserted myself and did not take the bait.

Just then the play came my way and I ran with the ball, away from where the men were standing. When I looked again, they were gone.

On my way home I learned that the day before a Russian army colonel was shot dead by a sniper at one of the busiest intersections in town. It was alleged that the shot came from the attic of a tall building opposite. The newspaper showed photos of the "shooter" in the attic, dead and burned, a rifle beside him. The whole scenario was bizarre and obviously fabricated. In those days, however, any legal challenge to that "evidence", no matter how eloquent or persuasive, would have been dismissed because it did not serve the purpose for which the whole scene was enacted.

The purpose, as we came to learn shortly thereafter, was to implicate our school in some alleged crime against the Russians who were the occupying power at the time. They accused two of our schoolmates of having killed a Russian soldier in a wood on the outskirts of town. The boys were forced to act out the "incident" in pictures published in a newspaper. They were tried and sent to the gallows.

Why single out our school in this manner? The Communists, who with Russian help held the reins of power, knew well that a school run by Benedictine monks was not likely to produce graduates devoted to their cause.

In the event, the school, along with other institutions run by monks, was later "nationalized", which meant that the monks were expelled and lay teachers took over. A few years later, when the Russians were no longer in occupation and a homegrown Communist government was firmly in place, our Head Master and several boys were arrested and accused of "armed conspiracy aimed at overthrowing the government". It was part of the program of dismantling elements in society not sympathetic to the party line. Don't bother with evidence, just bring down the verdict.........

More boys were sent to die. The rest, receiving 10 to 15-year prison sentences were freed from prison by the Hungarian Revolution of 1956. Even after that, for the next thirty years or more they were denied the privilege of working at jobs of their choice and could only qualify for menial work, as hospital porters or mortuary attendants. By then life and time had passed them by. They, their families and friends missed out on living in a way that can never be made good.

The lesson? Keep your eye on the ball when you play the game and ignore spectators on the touch line.

.......especially those wearing leather coats.

THE TRUSTED MAN

The word "neighbour" carries a friendly connotation. "Neighbourly" implies someone infused with good will, ready to stand by you in times of trouble, one who affirms your faith in the kindness, the integrity, the benevolence of mankind, one who inspires you to rush to his aid when misfortune becomes his turn.

Joe H. was none of that. He was our neighbour during the critical years when one dictatorship was followed by another. By rights I should have forgotten him long ago. Yet he made his mark, taught me a lesson which, perhaps as a means of survival, ever after coloured my assessment of strangers. I am not being unkind if I call him a rat.

Joe H. was a natty dresser, dapper in his double-breasted suit with a coloured handkerchief planted in the breast pocket. Maybe it is because of him that I have had a lifelong aversion to a bow tie. If he chose not to wear it, he would wear an open-necked shirt with a fluffed up scarf over his Adam's apple. A constant feature of his attire was a half-frozen smile, not one that hinted at benevolent amusement but rather that of a man who has just seen his worst enemy carried to the gallows. Invariably, he had a cigarette going.

He appeared to have no regular occupation. He voiced no opinions but was good at drawing opinions out of others. Religion? He had none, but given the right circumstances he would affirm allegiance to any religion that would fit in with the dictates of the conversation. He would parry inquiry into the minutiae of his religion of the moment by saying he was "not practising".

He was perfectly suited to an important role in the context of dictatorships: that of informer. He saw. He reported. He made a seamless transition in this role when one dictatorship was replaced by another. Friendly on the outside, ready to earn your sympathy, your confidence, your faux pas, letting slip whatever happened to be on your mind. By then we knew that the snake with the attractive markings is invariably deadly.

Within weeks of the establishment of Communist rule every house had a "trusted man" ("bizalmi") appointed by the authorities. This person was responsible for reporting to the "trusted man" overseeing the block any information, any gossip, any malevolent tidbit that would inform the authorities of the sympathies or otherwise of residents toward the government in power. This became a useful tool to the government when in the first post-war general election, touted as a "free" election, even despite the presence of the

Soviet forces, despite intimidation tactics, despite beatings administered by the Communists, the latter obtained only 17 percent of the votes cast. The "trusted men" went to work and with their help a large percentage of the population was simply removed from the list of voters, "disenfranchised", as the saying goes. Joe H. and his ilk had proved their usefulness.

Confident now, Joe H. went about his work, knowing that he had the upper hand. The apartment building in which we lived had no air-conditioning, so it was inevitable that on hot days we would keep windows and doors open to ensure a circulation of air. Joe H. sported the habit of enjoying the fresh air standing by our door, just out of sight but well within earshot of any conversation. Occasionally his cigarette smoke drifted in, betraying his presence. Now and then we would call out to him to let him know we knew he was there and ask him to move away because his smoke was bothersome. Occasionally we would surprise him from two sides as he leaned on the wall, eavesdropping. Embarrassment? To him it was no problem. Yet we avoided acute confrontation as it would only add to our risk. He could report anything he wished, factual or otherwise. If we were to harass him, he would find it cause for an adverse report. Any remark, any unguarded gesture was seized upon. A joke of the time featured three strangers sitting on a park bench. One heaved a sigh. The second spat out. The third reported them to the police for anti-government activity.

We learned to be wary. We agreed on plausible-sounding pseudonyms and bogus place names, a game which injected new interest into our conversation. We learned the usefulness of facial muscles as a means of silent communication. At times we lowered our voices, trying to draw Joe closer till his shoulder pads showed at the door.

These devices were amusing to a point but deadly serious. The stakes were high. We were vulnerable. Joe had no ethics, that much we knew, so our safety rested on his built-in stupidity or the tenuous efficacy of our games.

De mortuis nihil nisi bonum, goes the old saying: of the dead say nothing, if not good. I cannot say anything good about Joe H.

He did us harm, as we later learned.

NIGHT CROSSING

There was a time when the territory of Hungary extended to the Adriatic Sea. After the Trianon Peace Treaty of 1919 Hungary lost some 60 per cent of its territories and with it, its foothold on the Adriatic. The landlocked country that remained still retained a precious gem: the Lake Balaton. In the hearts of most Hungarians the very name "Balaton" conjures up an image of a blue, sunlit lake and endless summer joy. Our school had a summer resort on the eastern shore of the lake at a place called *Akarattya*. There we spent carefree days in summer vacations which our parents could not have been able to afford.

The waters of the lake join the Danube by way of the *Sió* (pronounced Shee-oh) River. It was at the mouth of that river that we had arranged a rendezvous with some of our friends who had made the journey by rowing boat all the way from Budapest. We met them at the lakeside town of *Siófok*. From there, the direct diagonal journey to *Akarattya* across the lake is about ten miles.

It was the night of the August full moon. We left *Siófok* at ten in the evening in two touring boats. Normally, a diagonal trip over such a distance would not be attempted, for the lake had its moods and our boats would not do well in high winds. Tonight was different. There was not a breath of wind. The lake was like a mirror, reflecting the full moon which was so bright that the stars paled in its glow. The lake was at peace. The temptation to cross was impossible to resist.

We settled down to a slow rowing rhythm. We were in no hurry: the longer the trip the better we liked it. Yet we knew we had to make it across before the wind woke up. We pulled on the oars slowly, letting the boat run with each stroke before beginning another. No force was applied. In the moonlight the contact of the oars with the water left little whirlpools which persisted in the silver reflection in our wake. The experience was so unexpected, so stunning, so beautiful that, instinctively, we dared not disturb the silence. The only sound: the boat cleaving the water, homeward bound. An air of unreality took us by surprise. This was a voyage never before encountered and, surely, never to be lived again.

Well out into the lake, far from the shore, someone began to sing, slowly, quietly, reverently. We all joined in. The song was "Old Man River". Since none of us spoke English, we sang in Hungarian:

"…..*és messze viszi a Misszisszippi*

dalom…"

Our voices took flight in the moonlit silence, over the quiet water. Our song was a hymn, a tribute, almost a prayer to a freedom we had never before felt, never imagined possible, maybe never would chance upon again, a gift from Life enveloped in our silent glide, embraced by the warmth of a summer night. We revelled in the journey across the still water with not a splash, intent on not waking, not disturbing the lake, which for a finite, all too short a while, was our world.

Because we did not know any other lyrics to the song, again and again we repeated the same passage, sang it over and over. The melody was so in tune with what we felt. The moonlight stayed in our wake all the way, a silver-lined path, an ethereal, once-in-a-lifetime vision.

"…..*és messze viszi a Misszisszippi*

dalom…"

We were all city boys, born and brought up in the bustle and clamour which is the heartbeat of any town, more so of the metropolis we called home. Tonight we were free from all that, the rules, the cautions, the conventions by which our young lives were regulated. We sang our song in a wide, open space which listened to us, accepted our voices and did not cast them back with an echo which would have told us there was a perimeter to our adventure.

I do not remember reaching the shore at our destination. I do not recall discussing this singular journey with the boys again. It still survives within me as a deeply personal experience, a vision of how beautiful life can be through simple choices made in a timely manner.

Not long afterwards the world turned upside down, the school, as we knew it, was closed, the summer resort and the boats confiscated.

None of that will erase the memory of our journey under the full moon and strains of our song fading into the darkness around us.

"…..*és messze viszi a Misszisszippi*

dalom…"

HOLIDAY FOR GREED

The principles governing Supply and Demand are well known. High demand commands higher prices while prices fall when demand is low. It is a basic tenet of economics. This reciprocal relationship has profound effect on our disposable incomes. Of course, it is natural to assume that the prices charged in the first place bear a reasonable relationship to the cost of the materials contained and the amount of work and skill required to make them saleable. We have made trust such a basic foundation of our day-to-day transactions for so long that we are disinclined to question whether our trust is well placed. Our complacency in this regard may have consequences contrary to our interests.

As for me, I have a jaundiced view of "supply and demand" because time and again in my life I have seen shortages become a holiday for greed.

I had my first brush with this mutation when I was barely fourteen. We had spent months under bombardment, artillery fire and in the crossfire of a house-to-house infantry battle, all of which exhausted our food supply. We were literally starving and I myself had advanced scurvy to the point where I was bed-ridden. Warfare was the only activity in the city, there was no transportation and shops had been closed for months. We were reduced to one meal of boiled beans a day, boiled without salt in the water because our salt was finished also. Still reasonably nourished when we sought shelter in the cellar a few months before, we emerged as walking skeletons. Hunger was with us all day, hour by hour, minute by minute, hammering at our consciousness, gnawing away at our insides, an urgent need unrelieved, made worse by the fact that there was no hope of redemption in sight. We were still alive but could not draw comfort from that because survival only prolonged our distress.

And then relief arrived! People from the countryside appeared, selling chickens and geese. If their faces were hard, their terms were even harder. Money had no value, so payment was in kind. The "kind" in which they were interested was gold jewellery. True, my mother had little by way of jewellery and such items as she did possess were family heirlooms whose sentimental value greatly surpassed the actual worth of the gold itself. The vendor handled the necklace with contempt and called it insufficient payment for the chicken in his basket. He wanted her wedding ring in addition.

Starvation plays games with your values. You are, or think you are, at the end of your road. The end of the road is death. If you turn the vendor away, you die, while still owning the treasures you were not willing to give away. What use are they to you then? On the other hand, if you buckle to the vendor's demands, you will enjoy a long overdue one-time feast and survive until the next hunger crisis pays a visit, by which time you will have less, or nothing at all, to bargain with.

The vendor was haughty, impatient. He was on the empowered side of the supply and demand equation. Our predicament was of interest to him only in so far as he stood to gain from it. That his gain was usurious and his price shameful bothered him not at all. He had no self-image to worry about. He was the predator, we were the prey. "Take it or leave it" was the choice he offered and he was in no mood to linger for a tardy answer. He got what he wanted.

In the ensuing weeks the same scenario repeated itself while we were hanging on to life. "Demand" was dictated by our hunger. Greed had a holiday. Everything of value that remained, clothing, bedding, items of comfort, was lost in this one-sided game of barter. Supply won out but my perspective of supply and demand was forever changed.

Tales of long ago, irrelevant to a world that has become more civilized in the almost 70 years that have passed? Think again.

In the Spring of 2010 the Icelandic volcano *Eyjafjallajökull* erupted violently, sending a vast cloud of volcanic ash across the face of Europe and parts of Asia. Airline schedules were disrupted. Holidaymakers, intent upon returning to their European destinations, were stranded in Australia and some Asian cities. This was a disaster of sorts for such travellers, particularly as many of them were at the end of their planned holiday and had already spent their proposed budget.

Again, there was an imbalance in the supply and demand equation. The planes were not flying, the passengers were stuck, all accommodation for those waiting to fly was filled. You would think that suppliers of hotel rooms would be only too happy that with all their rooms filled they were not losing money on vacancies. Full occupancy at the planned fair prices: a businessman's dream, right?

Not so. In many hotels in Asia and Australia hotel prices tripled, causing many travellers to sleep on the floor at airport concourses. Such airlines as were still flying to affected destinations, raised their fares, putting flights out of reach of the average traveller.

Supply and demand: think about it.

But then, we don't have to look far for examples such as these. Closer to home, in Canada, where everything is "oh, so above board", the price of gasoline is hiked during periods of peak travel, such as summer holidays, long weekends and the Christmas celebrations. These increases in gasoline prices have nothing to do with variations in world prices of crude oil.

The market: does it deserve our trust?

Clearly, the weigh scale of supply and demand needs an ethical fulcrum. In its present state it gives the business world a black eye. In times past when the buyer stood face to face with the vendor he had a chance to upbraid him if the price was exorbitant.

The world has changed: the individual buyer is still on his own but the vendor is now largely represented by big business against which the lowly customer has little chance of redress. The result is frustration, apathy. We throw in the towel, capitulate to unseen forces behind the scenes and suffer manipulation of prices to the point where we are held to ransom. Even the government, beneficiary of increasing tax revenue from escalating gasoline prices, turns a blind eye to the public's distress, stays mum to protect its own interests.

Supply and demand: without an ethical component it can be rigged to become a holiday for greed.

THE CRUCIBLE

It was well after nine, almost ten, in the evening when the doorbell rang on November 3, 1947. At the door, a close family friend explained in hurried whispers that I was to leave for England the day after tomorrow. He said I should not advertise the fact, for it was only by a piece of luck that he managed to arrange a flight through Prague and that loophole to leave Russian-occupied Hungary would not remain open for long. I do not remember seeing him leave, I just stood there with five English pounds and an air ticket in my hand, as if hit by a tidal wave which would carry me I did not know where.

True, I had been chosen to go to Downside School in England on a year's scholarship some months before but a long delay in getting a passport and failure to receive a permit to leave Russian- occupied Hungary had me already miss almost two months of the school year and I had all but given up hope of ever going there. I continued to attend school at home but the uncertainty left me half-hearted about my studies. In the perilous circumstances then prevailing returning to the school to take leave of my friends was out of the question.

The simple routine of our daily existence was suddenly dislocated. We had just one day to get me ready to travel.

I had no suitcase, no clothes and we had no money to buy them. I don't think any of us slept that night. Faced with the sudden prospect of losing their only child my parents nevertheless knew that they had to do what they could to provide me with the wherewithal which a young boy may need when he goes away for a year. It was a time for quick action. There will be time for tears later.

As for me, I could not picture an existence away from my family. Our circumstances had never allowed travel beyond short trips in a third class compartment on the train. While I was excited at the prospect of a major journey, by airplane at that, to a distant country whose language I did not speak, I felt a shiver of insecurity about days to come for which my imagination had not prepared me. At home the love and support of my family was assuredly mine and it did not occur to me that within a few hours I would place between us a void filled with minefields, guns and vast distance which in my impoverished state I could not hope to traverse. I was already on a fast-moving train and I could not get off. Whatever the coming days would throw at me I would have to face on my own. I was unprepared.

Hours passed like minutes. My grandmother, a pivotal figure of my childhood, sat silent. She knew she would not see me again, that this was the parting of which she had spoken and which she feared, yet she must not interpose her own feelings into the process that was already under way: my life's journey which was about to begin. There are good-byes in life which are more difficult than dying for there is no natural end to stop the pain. Looking at her I saw her back in my childhood, bending over my pillow for a good-night kiss, then singing to me till I fell asleep. I remembered our long afternoon walks along the river, while my parents, who both worked night shifts, slept during the day. Too, I recalled her courage in the war when in a small village bungalow we sat under a table with bombs falling about us and machine-gun bullets stirring up dust in the courtyard.

In the afternoon my father returned from the flea market with a cardboard suitcase, a second-hand suit, shiny from wear, with a highly obvious patch on the back. He bought a pair of well-used brown brogue shoes with a hole in the sole which the cobbler quickly covered with a patch of leather. Then there was a barely credible jacket, converted from what had been a German army officer's tunic. Did he die wearing it? I wondered. Together with three shirts, underwear and a few pairs of socks and a tie these constituted my wardrobe for travel. In marginal existence there is no room for fashion.

Meanwhile, the clock kept ticking.

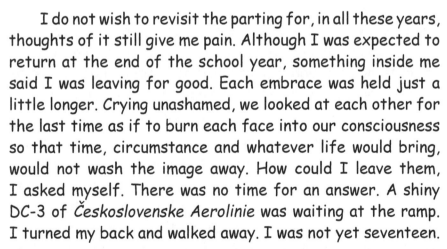

I do not wish to revisit the parting for, in all these years, thoughts of it still give me pain. Although I was expected to return at the end of the school year, something inside me said I was leaving for good. Each embrace was held just a little longer. Crying unashamed, we looked at each other for the last time as if to burn each face into our consciousness so that time, circumstance and whatever life would bring, would not wash the image away. How could I leave them, I asked myself. There was no time for an answer. A shiny DC-3 of Československe Aerolinie was waiting at the ramp. I turned my back and walked away. I was not yet seventeen.

Of the journey I recall little. Prague, though drab as every post-war city, was still beautiful. I slept in soft sheets in a little hotel at Resnicka 19. Everyone was kind, except the tramcar conductor who angrily made me get off because I tried to pay for a small fare with a banknote.

London was blanketed by fog so dense when I arrived that it took three hours to reach town from the airport. Someone kindly made a hotel booking for me but the taxi driver got lost in the fog. I ended up at the *South Kensington Hotel* in Gloucester Road where the rate was fifty two shillings for the night, eating up much of my money. I dined on the remains of my *szalonna* and crusty bed and after a bath in the deepest bath tub I had seen I fell into deep sleep.

On arrival I quickly learned that my English was totally inadequate for survival, no matter how kind the people I had met thus far. A strange, somewhat intimidating new experience.

Daylight brought new sounds into the street below, so different from the clamour of the road at home. From my high window I looked out on the rooftops at hundreds of chimneys, each a question-mark, probing the days ahead. I was immensely homesick but there was no consoling word, no understanding touch. Unprepared as I was, I had to live by my own counsel from now on. I had better toughen up!

A kind family gave me lunch and put me on the train to Bath, Somerset. We arrived in the dark. It was raining heavily as I approached a railway man in uniform to ask for directions. In truth, all I could say was "Downside School" and he pointed me to a Green Line bus stop in the street outside. The bus conductor realized that "Downside School" was the only useful English I knew and he indicated that he would tell me when to get off. It was dark along the route.

The bus made so many stops that I had reason to believe we had gone past the school. Eventually the conductor's gesture indicated that I had to get off at the next stop. As I stepped off the bus in the village of Stratton-on-the-Fosse the notion that I was alone hit me hard. From now on everything was to be a challenge, a measure of my ability to adapt, cope and survive, not least because my English was virtually non-existent.

A modest sign on the high stone wall read "*Downside School*". The open gate led to a curving drive under large trees. The patter of the rain on the leaves filtered all other sound. I had to switch my heavy suitcase from one hand to the other a number of times. In a nearby building all the lights were turned on, so I headed in that direction. There were lively sounds within and it was no surprise that my knock on the door went unheard. Hesitantly I opened the door. Men and women were having a good time in an aura of beer and cigarette smoke. My arrival seemed to amuse them and all eyes turned to me.

"Downside School", I said hesitantly. This brought forth a burst of laughter from the merrymakers but one man kindly took me outside and around the house where the imposing complex of the school buildings came into view. Across the wide yard and the rain the murmur of the school reached my ears. I had arrived at my destination. It was 8.30 in the evening. Whatever will happen will happen now.

A heavy swing door led into a hall with leaded bay windows and an arched ceiling. Boys in black jackets, white shirts and striped grey trousers scurried to and fro and the air was heavy with the smell of wood polish. A blacked-robed monk stood facing me, a large man, squinting at me through strong glasses. His expression was severe but he shook my hand and took my suitcase. It was almost as if he had been expecting me and in retrospect he probably was, because the station master in Bath kept the Head Master abreast of the movements of boys connected with the school. I told him my name and he bade me to follow. The office door bore the sign *Head Master* and only then did I learn that I had met Father Wilfrid Passmore, Head Master of Downside, a man who was to have a profound influence on my life, who was to become a father figure, a mentor and an inspirer in the days to come. To be sure, he was a hard taskmaster but I learned that later.

With the remarkable clarity of gestures, emphasis and words possible in a language I did not yet understand he explained that I had arrived so late that there was no room available for me at the school. Still, I was welcome and, not to worry, I would learn English soon. He grabbed my suitcase and led me through corridors and stairways to what was to be my room.

At the top of the stairs we could go no higher. He turned into a linen room, into which by now an iron bedstead had been placed, along with a desk, a chair and some towels. He apologized for the accommodation and told me he would come for me early the next day. There was a bathroom on the same level. I had a quick bath and got into bed. The room was cold and the blankets were inadequate but in the end my fatigue dictated sleep and I did not wake until morning. By the time Father Wilfrid came for me I was ready.

I was conducted to the refectory where the boys, some four hundred of them, were already having breakfast. He assigned a seat to me at a long table and told me he would come for me after the meal. At the table I was surrounded by curious glances. Some of the boys tried to talk to me but

I shrugged my lack of comprehension. A few found my lack of English amusing but most of them soon lost interest and went on with their meal. I was delighted to have not only porridge but also egg and toast, luxury items to me after three years of starving.

In his office the Head Master handed me a sheet of paper with all the lecture periods of the week. I was astonished to find that he assigned me to every period, every school day, from 8.30 to noon and from 1 p.m. to 5 p.m., along with a few evening classes. He explained that I was there to learn English and I could only learn it if I was surrounded by the spoken language all day. Clearly he was right. I was desperate to have time to myself to contemplate my homesickness but that was not the objective of my long journey here.

Physics, mathematics, chemistry, Latin and English literature were all on my menu and I dutifully attended all my classes. At first I felt I would never comprehend what was spoken but daily some words and phrases began to sound familiar. Strangely, mathematics was the easiest for the equations written on the blackboard represented a universal language with which I was already familiar. In fact, with quadratic equations I always had the answer ready first because the Hungarian way of going about it was quicker. My teacher appreciated being introduced to a new approach. In each class the teacher would frequently ask if I could follow. It was slow at first but I knew I was making progress.

Physics has its own language but the teacher was remarkably kind and helped me along. The chemistry teacher, on the other hand, mocked something that I had said and invited the whole class to join in the merriment. I was not going to put up with that and told the Head Master that I would not attend chemistry classes again. He understood.

Latin was a joy, because I had had six years of it in Hungary and it was all familiar. Translating it would have to wait.

The class in English was working on Chaucer's *Canterbury Tales*. "The Nun's Priest's Tale" gave me hope because, although it was Old English, the words were similar to German and very quickly I was able to get up to speed. It was in English that I felt I made the fastest progress and I was grateful to Father Wilfrid for his foresight and wisdom to throw me into the deep end and let me swim as best I could.

At the dining room table I began to understand more and more of what the boys were saying. I just had to write off the fact that my mistakes were funny for some.

Winter weather arrived in late November and my "room" was cold. I sat close to the lukewarm heater on the wall while doing my assigned homework. Although I was there only to learn English I had already decided that I would catch up with the curriculum. In the evenings I looked up words that had caught my attention. I even made a pencil dot along the margin if I needed to find the meaning of new words in a book. This became a tiresome exercise and then I just read on, telling myself that the meaning would become obvious if I just swam with the current. I was right.

The laundry was picked up on Thursdays. Every item we wore was marked with black ink, showing the surname and the school number, which in my case was 131. I put two shirts in the wash, thinking they would be back the next day. Little did I know that the laundry took a week to return. I had no option but to wash my remaining shirt in the bathtub and hang it up to dry. In the cold, damp surroundings it was still wet in the morning, so I had to wear it as it was. Worse was to come. With three square meals a day I began to put on weight and all my shirts became tight at the neck. There was nothing I could do, so I wore them wet and tight for several months, the collar strangling me all day, for I had to wear a tie.

The patch on the back of my shiny suit, although I could not see it, felt like a burning ulcer on my back. The dead German's jacket remained unused, it was hardly an item to wear in my current setting. We had to attend church most mornings and four times on Sundays. I dreaded having to kneel because it revealed the patch on the sole of my shoe. By and large, however, no one made any comment on my attire.

In fact, the boys, all of whom came from well-to-do backgrounds, let me fit in easily and ignored the patch on my jacket. They judged me for what I was. Only one boy, one I did not know, ever made an offensive gesture. I was standing by a sports field, watching a game. As he walked by me he deliberately stepped on my toe without uttering a word in regret. My heart said "knock his teeth out"! My head said "let it go, you are a guest here"! I let it go. I did the right thing.

The breakthrough arrived in my second term. Suddenly I felt a facility in the English language, it was almost like learning to ride a bicycle when from one moment to the next you have found your balance. The Head Master had me measured for a uniform. I still remember the joy of wearing it for the first time, the joy of being dressed like every other boy, the relief of no longer parading the patch on my back. My shirts still strangled me at the neck and it was some time before I got new ones.

I spent the month-long school holidays at a small seaside hotel in Ramsgate, on the Kent coast. My experience there is described elsewhere. By this time I had learned to live with my homesickness, secure in the knowledge that at the end of the school year I would be heading home.

As for my English, it took me about six weeks to begin to make basic conversation. After three months I was reasonably fluent. After six months my teacher read out my English essays to the class as an example of how to write literary criticism. This was all the more astonishing because I could not have done it in Hungarian. I felt an affinity for the English language, denied to me in my native tongue. The poetry of Browning and the lavish imagery of Shakespeare held me spellbound.

With the arrival of the summer term I received the new, grey uniform worn at the school at that time of year. My joy was short-lived. At breakfast the waitress dropped a plateful of scrambled eggs on my jacket the first time I wore it. By that time I was so inured to setbacks that I did not lose sleep over the incident. My studies were going well and I was receiving encouragement from all sides.

Except for field hockey, I played all the games at the school. I still remember running with the ball at rugby, being tackled around the ankles, sliding along the wet grass, inhaling its exhilarating scent. Soccer was not an official game at the school but I was on the school team, playing against local village teams. Because I distinguished myself as a goalkeeper in soccer I was conscripted to become wicketkeeper for my house cricket team, never having seen a game of cricket. After some lessons in the squash court the night before, I walked on the field in my new role the next day. When it was my turn to bat I remembered being told to swing at the ball and try to score some runs. It was all too new, too fast. I stopped the first ball I received. I swung at the second and made contact, but forgot to run. As it happened, there was no need to run, because I had hit it for a six, the cricket equivalent of a "home run".

It was during the summer term that I received a letter postmarked in Oklahoma. It contained a scribbled note in my father's handwriting, advising me not to return to Hungary under any circumstances. The political situation had become grave, and as I was to learn later, my Head Master, my cousin and several of my immediate friends at school suffered immeasurable hardships under the ruthless regime in power. I chronicled the subsequent developments elsewhere in this volume.

My summer holiday by the sea brought me my first romance. I fell, head over heels, in love with the Belgian nanny who cared for the owner's children. It was all too innocent, too beautiful and too sad, for we both knew that it had to end. When she left for home I accompanied her on the bus to Dover. "One single and one return to Dover" I called out to the conductor. As he handed me the tickets in a stentorian voice he repeated "one single and one return". The voice of Fate shutting the door on our idyll. I returned to school with a broken heart.

During my second year I was given a fine room, overlooking a lawn, presided over by proud old trees with generous foliage. Studying in these surroundings became much more pleasant. Any anxiety I may have had about not measuring up was soon dispelled because I came to realize that I was as good as anyone in the class. When I left the school the Head Master thanked me for keeping my room tidy. He said he always showed my room to parents of prospective students.

Stephen came into the picture at this time. He saved my life by making it possible for me to stay in England. More about him later. The school accepted me for a second year, this time as a fully- fledged student, a candidate for the "Oxford and Cambridge Higher School Certificate". I worked with added incentive and received high marks, especially in English and Latin. Mr. Robin Athill, my English teacher gave me wings. A young boy needs acknowledgement and an occasional pat on the back and he rewarded my hard work with both. As it happened, I passed the examination easily and was even made to write the exam for Distinction in English. I fell just short of the mark required for a scholarship. In truth, I would not have had the funds to follow on to Oxford or Cambridge. A much tougher alternative awaited me in London.

At prizegiving I received the prize in English. Father Wilfrid handed it to me with a glint in his eye and a warm handshake. His severe timetable had done its job. My friend Michael Binns' mother told me that she had applauded me for my mother who could not be there. The Binns were good to me. On their visits to their son they always invited me along for dinner in some nearby town and were gracious hosts. I kept in touch with Michael after I left school but once in the grip of my brutal schedule in London I lost touch with all my friends.

My other friend at Downside was Michael Woodward, a boy from Chile. He, too, was most generous to me. During one of our holidays he invited me to stay with him, his brother and sister at Godstone, Surrey and gave me a carefree week of sport and fun. We stayed in touch in years to follow. He returned to Chile and worked as a parish priest, devoting his life to the poor. When the Pinochet coup seized power in Chile Michael was arrested and died aboard the ship *La Esmeralda* from injuries received during his interrogation.

Because of his dual English/Chilean citizenship his death became a *cause célèbre*, even a matter of debate in the British Parliament. The family tried to get his body for burial but was refused. The only person who knew where Michael was buried was afraid to talk. It is said that a highway complex was built over the site of interment, so the remains will never be found.

At the risk of being anachronistic, I must mention that in 2005 I chanced to meet an engaging middle-aged man on the beach in Bali, Indonesia. He was attending a conference at the next door hotel, representing the government of Chile. He told me he was from Valparaiso, the seaport where Michael died. I told him I was glad to have the chance to talk to him and mentioned that my best friend had died aboard *La Esmeralda*. I mentioned Michael by name.

At this the man's demeanour changed instantly. He became evasive, reluctant to talk. Finally he admitted that he had heard "something about a priest" but really had no knowledge of the matter. We parted on friendly terms but much sooner than I had anticipated. I dropped off a letter inviting him to dinner but did not receive a response.

My scholarship to Downside, my two years there, were a life-changing experience for me. I was lucky to get there in the first place. The Benedictine monks both in Hungary and in England gave me a chance to make something of my life. They asked for nothing in return. All I could do to thank them was to work hard and justify their trust. That I did. The work ethic I acquired has served me well. I now know that all the good things that came to me in life were the result of periods of intense hard work.

―――⌘―――

OLD SALT

After a lifetime at sea Surgeon-Commander W.H. Austin Sinclair-Loutit, R.N. retired to a small seaside hotel on the Kent coast. He never married. He had a family of three: an upright piano, a stately Daimler car and Brock, his faithful bull terrier dog.

    He occupied a fine room in the hotel, overlooking Pegwell Bay and the Goodwin Sands, with the coastline at Deal discernible across the bay on clear days as it curved away to Dover. He spent much of his time in his favourite armchair, pipe in hand, looking out to sea and at such times he seemed to be sailing some distant ocean, making landfall in some far-off port with an exotic name, perhaps in the arms of a girl who bade him stay only because she knew his ship would sail in the morning.

At times he would sit at his piano, playing his favourite Chopin nocturne, revisiting the same passage time and again, his thoughts a world away. He would ask over his shoulder:

"My dear Nicholas, do you know what these notes say?"

I shook my head. He kept on playing.

"……no,……no……nevermore…", he sighed, "…that's what they say……".

Austin was an Old Boy and benefactor of my boarding school in Somerset. When the term ended I faced a month's vacation which I could not spend in an empty school. My Head Master arranged for me to stay in the hotel, with Austin keeping a kindly eye out for me until I was ready to return to school for another term. As it happened, thereafter I spent all my school vacations there and Austin turned from a mentor into a friend. He saw to it that I went to St. Augustine's Church on Sundays, where the Belgian priest, Father Odilo, in a sleepy monotone, invariably announced: "……mass daily during the week at quawter to seven, 'alf past seven and quawter to nine……".

Now and then Austin would phone. "My dear Nicholas, would you care to accompany Brock and myself for a walk on the Promenade?" I always agreed. "Well then, come over at eleven thirty for a drink and a smoke". He poured generous helpings of Australian muscatel and we would converse through clouds of smoke from the fine, thick Egyptian cigarettes, tailor-made for him by Benson and Hedges. Brock would lie by his feet and pretend to sleep while actually listening to the conversation.

Brock was a fine bull terrier, the son of a Champion of Champions at Cruft's, the premier dog show of the day in England. Austin spent time explaining with pride the refinements exhibited by his friend and the dog and I took to each other quickly. I appreciated his stance, his gait, the way he pricked up his ears, his singular attention when he was addressed and his uncanny ability to act on words of command even when they were concealed in a sentence during conversation. Clearly, the old man was proud of his companion and I had to admit I felt privileged in their company.

The Promenade was frequented by little old ladies, mostly widows who lived up at The Terrace in stately apartments giving on to the sea. An air of gentility prevailed. Everyone deferred to Austin as "Commander" or "Doctor". It was an education hearing him exchange words with a passer-by without stopping and yet making the encounter seem meaningful. The smell of the sea enveloped us all, it bestowed on us a sense of belonging, a reference point, a notion that somehow we, too, were sailors. That salty ambience is still fresh in the memory after the lapse of decades.

Austin regaled me with tales of the sea and of sailors long gone. He introduced me to understanding the way an Englishman thinks, and, although he did not warn me, at once I knew that it would be a lifetime study. He let me in on some raunchy sailor humour in prose, verse and song. Looking out to sea I no longer felt confined within the narrow perimeter of my childhood. Out beyond the mists and the waves there beckoned a freedom, an invitation, a challenge, an excitement which, as a newcomer to England, I could anticipate but not yet picture in my imagination. The sea was life itself and life would take me anywhere I cared, or dared, to go.

Once we reached The Promenade Brock's genteel behaviour underwent a transformation for which I was not prepared. By this time Austin was in his late sixties and his light frame failed him against the powerful pull of his dog. On arrival in this arena Brock was quick to size up the opposition from afar. As unsuspecting old ladies drew near, parading their equally unsuspecting pomeranians, pekinese and pugs Brock's primal instinct switched into the attack mode. The hapless women tried desperately to pull their mutts away from harm while Brock was pulling Austin, much too weak to stop him, toward the prey. The dogs yelped and barked and the fur flew amid helpless cries of "really, Commander!!!" and "dreadfully sorry, Mrs. Deveson....!!!", all the while leaving a trail of blood and fur along the Promenade which guided us the back to the hotel.

Despite the skirmishes everyone remained on good terms, recognizing the fact that there was a side of human life where the animals had undisputed control. Thus, Austin was frequently invited to tea parties and soirées at The Terrace, where the old ladies lived. They made sure that I was invited along with Austin and concluded the invitation with an apologetic "......of course, you understand......" which meant that Brock was persona non grata in their homes.

The tea parties were time borrowed from a past age. The paper-thin cucumber sandwiches were dainty and delicious, the tea pot warmly embraced by a loving tea cozy, the cups and plates sculpted of fine china, the silver teaspoons and strainers polished to a flawless shine and each setting attended to by a monogrammed napkin. The conversation was charming, if not profound and everyone took pains to say something complimentary about my fledgling English. I liked the atmosphere immensely and marvelled at how these people came out of the war, only recently ended, so unscathed while I was still nursing deep scars.

The soirées invariably featured music and singing. Someone would sit down at the piano to play "Old Father Thames" or "A Perfect Day" and there was always a man willing to belt out the lyrics. The applause was spontaneous and heartfelt; sincere, ordinary people enjoying a moment of their culture, their way of life. All this came so naturally to them, they seemed untouched by the rancour and hate of which I had seen so much during the war.

This is a cherished picture of England, of a generation alive only in the haze of memory, a wistful reminiscence which visits me now and then as it has for decades.

In the hotel dining room there was no conversation. Austin, like the captain of a ship, took his meals alone. When guests were few the silence was oppressive. If a diner dropped a processed pea back on his plate, everyone knew. The waiter went about his duties without uttering a sound. He had thin, grey hair brushed straight back, skin discoloured from years of smoking and eyes of an intensity which, if only focused on some worthwhile endeavour, might have made him a man of renown. Instead, he bore a sinister air, redeemed only in part by his polite manner, appropriate to his occupation. Still, I always suspected that his obsequious smile as he approached was supplanted by a smirk as he turned and walked away.

On my next visit to the hotel a surprise awaited me. We again enjoyed the ritual Australian muscatel and Egyptian cigarettes before Brock's morning walk. I still had a modicum of apprehension as we approached the Promenade where a sprinkling of old ladies and their dogs was already enjoying the sea air. However, this time Brock was on his best behaviour. "Good morning, Mrs. Deveson!". "Good morning, Commander!". All smiles. No growling, no aggressive posturing, no attack.

It was all you could do to stop Brock from saying "how do you do" to the old lady and her dog. At each ensuing encounter the same civility prevailed.

I could only keep quiet for so long. Clearly, Austin had conquered the aggressive traits of his dog. It must have taken much of his time and complex reasoning with this highly intelligent animal. The result was astonishing.

"Austin, I bow to your skills. You achieved wonders with Brock. You have turned him into an amiable, courteous, gentle creature!"

"My dear Nicholas, you are not very observant! Else you would have noticed me filling Brock's bowl with Australian muscatel before we left".

MY HOME, MY CASTLE

The many London addresses of my youth had one thing in common: they made me feel like a prisoner. I never felt at home. I cannot recall one chair, one spot, one niche where I felt at ease, a place to regroup when I needed comfort. They kept out the rain but not the cold. On chilly nights the gas heater gobbled up the shilling coins one after the other, signalling its appetite for more as the flame went out with a "plop-plop- plop". The ill-fitting windows gave free passage to drafts and when the wind blew gusting currents of air wailed mournfully, as if through the rigging in a storm at sea. Even in moments of peace I longed to move to a better place but it remained only a wish: I could not afford it.

When you are poor you have little choice in where you live. The London of the nineteen-fifties taught me that. Once gracious dwellings were divided up into a number of one-room bed-sitters, the shape and size not designed with the occupant in mind. All the residents of the building shared one bathroom and one telephone, sometimes two floors below. A single gas ring on the floor served as a kitchen stove. The view from the window mirrored what was seen from across the street: nameless figures in windows, all swimming against life's current, like myself.

After two years spent at an expensive boarding school in the country, taking up residence at 119, Grosvenor Road, Pimlico, was a harsh plunge into reality. The small house was bulging at the seams with tenants, mostly in their early twenties, all of them down at heel and laconic to boot. On our way to God knows where in life, we were there for shelter and food, not to talk to each other. I occupied the front room, complete with a bay window and a convincing view of the tall chimney stacks of the Battersea Power Station across the Thames. The room had a feeling of never having been lived in. I even had trouble deciding where to sit when I just needed a place to read. Right from the start I was reluctant to unpack because the place did not feel like a destination reached.

A condition of my rental was allowing the landlady's daughter to take her piano lesson every Tuesday on the upright piano in the corner of the room. The male piano teacher came to the house and was working with her on the "Blue Danube" waltz which she played with the rhythm of wounded soldiers marching while wearing full length plaster casts.

Breakfast was served at seven sharp. The landlady worked to a schedule and did not tolerate tardiness. We sat on long benches along a table. I soon learned to ease myself in east to west, for trying in the opposite direction a sharp split in the wood could inflict a nasty stab on the buttock, such as a soldier might sustain while running from a bayonet charge. Lumpy porridge came first, followed by toast. The slices looked like funeral notices, burnt at the edge. There was no conversation. The butter and the marmalade were passed in response to signals. The marmalade jar stopped in front of me one morning with a dead fly, belly up, embalmed in it, an inch down. The boys were not bothered. The jar continued its rounds.

An exception to the quiet, sullen bunch that we represented was Doni, a student visiting from Switzerland. He was the landlady's darling. He could come as late as he wished and still remain in favour. One morning he appeared with a map in his hand and loudly announced the itinerary for his forthcoming trip through the British Isles. "I shall kiss the Blarney stone! ", he declared, with enthusiasm.

"Kiss my arse!" whispered a voice from the bench.

It was a blessing when, not much later, I moved in with Laci and his wife Rose at 35, Gratton Road, Hammersmith. They rented the second floor in a dilapidated house and sublet the attic to me. Right behind the Olympia Exhibition Hall, the area had been heavily bombed during the war and the building was surrounded by bombed sites, rubble still piled high. There was no street lighting. It felt like being back in the blackout.

Three things made this place memorable. By far the worst was the sight of a morbidly obese woman in the building opposite who performed sluggish callisthenics in the nude by the open window. The family occupying the ground floor apartment downstairs always cooked mutton on Sunday mornings the smell of which was particularly difficult to endure for the rest of the day. Then there was the antiquated hot water heater in the bathroom which required a steam engineer's ticket to operate.

I had trouble working the hot water system. One morning, while in my bath, the door burst open and the man from the ground floor entered, screaming obscenities at me, claiming I had flooded his kitchen below. He pulled up the sash window and threatened to throw me out. Laci appeared, holding a soda water bottle which he promised to break over the man's head if he touched me. A reluctant cease-fire ensued. As the man retreated he told me he knew I got home late every night, so he would wait for me among the ruins and do me in. Thereafter I always walked in the middle of the street, hoping for a chance of early warning.

Laci earned his living as a handyman. He had a purpose in life. One day he came to see me, all smiles. "Chief, we are moving! I have bought a house!" And so we did move, to 22, Lonsdale Road in Notting Hill. The house hid in the middle of a row of attached buildings, all looking exactly the same. I had a small room which received the sun all day. There was a drawback: the house had no electricity and no hot water.

For several months before electricity was installed I had to study by the light of a petroleum lamp. Eventually Laci set up a gas water heater in the bathroom and, at long last, we were able to wash in warm water.

This came at a price. One morning I was just finishing my ablutions when I suddenly felt dizzy and dropped my towel. As I picked it up, my arm began to jerk repeatedly. On the point of passing out I realized there must be a gas leak and I made it to the window to unlatch it. Laci found me unconscious some time later and carried me up to my room. I had a searing headache the rest of the day.

The Kensington borough newspapers, The News and The Post appeared once a week, on Thursday, carrying hundreds of advertisements for accommodation, bringing the hope of something better. The problem was, I got home from my night classes much too late, so I could not even begin to search until the weekend, by which time the good rooms had all been taken. Even then it meant a lot of foot-slogging going back and forth only to find that the landlady was a witch, the smell in the house repellent, the room dingy, the bath unclean. So back to the old "digs" for another spell.

I chanced to meet a man who told me his tenant was leaving. I could rent a bed-sitting room with a fireplace, pantry and use of the kitchen. Since they were away much of the time, I could have the place to myself. This was too good to pass up and I soon became a resident at No.2, St. Luke's Road, Westbourne Park, on the Metropolitan Line. Not my favourite area but the accommodation promised to be the most convenient I have had. I moved in, filled the pantry with food and felt elated at my new comforts. Late in the afternoon I prepared my supper, laid my table opposite the fireplace and set out the food. I was a bit put out by the fact that the landlady, surely, it must have been the landlady, had been into my butter with a serrated knife, but sharing food with others never was a problem for me. I ate heartily and then made tea in the pot from the mantelpiece. My life had taken a turn for the better, I knew.

A feeling of being "at home" pervaded me at last as I poured tea into the cup. No longer the prisoner of the gas ring, I could now make tea in "my own" kitchen. Three years of cold, draughty rooms, bombed sites all around, no electricity, no hot water, leaking gas lines, threatening neighbours, they were all a thing of the past. Time to relax, time to enjoy what I now had.

I raised the cup and took a sip. At once I was repelled by the pungent smell of stale urine! In disbelief, I took the lid off the teapot, only to unleash a hot blast of the same. Someone had used the teapot as a urinal and let the urine evaporate, leaving the concentrated smell of its contents behind. In revulsion I poured the tea down the toilet and got rid of the pot. In all my life I had not experienced anything so revolting, not through the war, not through the hardships that followed it and not in the three difficult years in London. Still, I thought, after washing my mouth out a dozen times, I must dwell on all the comforts that come with this place and not let my experience with the teapot spoil it. I had taken many knocks in life, this was just another.

My little radio kept me company. It brought into my room a world beyond my reach because my finances compelled me to stay within a well-defined perimeter. I listened to radio drama, then some music as daylight faded and in the end concluded that I had had a good day. I laid my head on the pillow and thought my day had ended.

As it turned out, not quite.

Within seconds of my putting out the light, scratching noises began under my bed. Strange, I thought, and switched the light back on again. The noise stopped abruptly. Imagining things, I thought, and put the light out again. The noises returned. They criss-crossed the room and would not cease. I turned the light back on again and got out of bed. I looked under the bed and found two mice looking at me. I heard noises from under the wardrobe and when I looked under that I saw several mice frozen in anticipation of what I might do next.

Wide awake and suddenly under siege by mice my new-found contentment left me in a hurry. I had never had an encounter with vermin like that and, not knowing what I should do next, I got my tennis racquet and tried, in vain, to swipe at the elusive intruders. I made no impression on them. Finally I gave up and returned to bed, noise and all. Eventually I fell into troubled sleep.

The next morning I entered the kitchen only to find that it was overrun by mice. There must have been forty or more of them all over the floor, running along a wooden strip around the wall and I could hear them in the long curtain. I shook the curtain and three mice dropped to the floor, scampering away towards a hole in the wall at floor level which was blocked by a fat mouse which was unable to move either in or out. I swiped at a passing mouse with the broom and must have broken its leg because it sped away with a pronounced limp.

And then it dawned on me that it was not my landlady's serrated knife but the incisor teeth of mice which had helped themselves to my slab of butter the day before. I recoiled at the thought of having eaten some of it. In fact, I discovered a large hole in the pantry wall through which the mice gained free access to my food.

I was horrified. I discarded all my uneaten food and began eating out in tea-shops, unable to face another meal in the house. Immediately I decided to move elsewhere but with my late hours and the difficulty in finding alternative accommodation I had to spend several more weeks in the company of mice, brazenly present both day and night, as if they knew they had dominance by sheer numbers. It was a nightmare which robbed me of sleep at night and continued into day with mice scampering to and fro, mockingly reminding me that I was powerless to do anything.

Fortune smiled on me because my next digs at 22, Gordon Place, Campden Hill, turned out to be the best I had so far. I had a view looking out over some trees and the sun visited my window in the afternoon. Back to the old gas ring but this time I had a wash basin in the room and the bathroom and telephone were a short flight of stairs away. A good place to return to at the end of the day. The Tempels, an elderly Jewish couple, refugees from Poland, ran a model rooming house and enjoyed the collective love and respect of their tenants. They gave us all a home.

The telephone was located on the first landing. The phone would ring in the Tempels' basement apartment. They would go to the bottom of the stairs and in fine voice call out the name of the tenant wanted by the caller. On the top floor lived a Miss Kandt, a refugee from Germany, the recipient of many telephone calls. Mr. Temple, fluent in

German, would call her name as it is pronounced in German, much to the amusement of the male residents.

This idyllic state of affairs continued until the Billiks, an old Russian couple, moved into the apartment below. I arrived home from my evening classes around 10.30 four nights a week and then had to prepare food before I could study. Although I moved around the room quietly, the creaking of the floor disturbed the Billiks below. One night I was startled by a savage beating on my door. When I opened it I found Mr. Billik in a rage, yelling "you are murrrrrdering my vife!". Soon after this incident I received notice "to quit and vacate" my room within a week.

The memorable event about my next address at 17, Harrington Gardens, Kensington, was the murder of the night porter in the hotel next door. Miss Belman, my landlady, like many Britishers, was fascinated by the mystery surrounding this occurrence, more particularly because the police could not find useful evidence on the crime scene and there appeared no motive for the heinous act. A cloud descended on the street and everyone looked upon everyone else as a possible murderer.

For a while in the wake of the murder Miss Belman, acquainted with my movements, would waylay me by the front door and elaborate upon "her theory" about the crime, with daily updates as time passed. Her reasoning was original and riveting, far beyond the coordinates of my imagination. She must have been a student of Agatha Christie, if not a crime writer herself. Her theories notwithstanding, the culprit was never caught.

The hardships endured in these sundry places worked towards a useful end. They were a spur to serious study. Accommodation, no matter how I longed for better, was a peripheral, not a central issue along the way. I knew that the only way to escape this environment was by my own effort, passing through the minefield of endless examinations. There was no time for diversions, parties or even just living. If I did not falter, and only then, would I finally emerge into the light. It was like climbing a mountain. At first you toil, tread and slog through trees and brush, your sweat the only reminder of your uphill load. The peak is not in view. For all you know, it may not reveal itself till you get there. It throws fatigue, pain and doubt at you to blunt your will. You must believe it is there, just to make it from one step to the next. The words "either, or" challenge, tempt and tease you time and again. Either you press on, ignoring pain, thirst and exhaustion or turn from looking skyward and descend in defeat, an also-ran, back to the old grind, with your hopes and chances clothed once more in the dust of the road. Competence is a pedestrian, everyday obligation for survival. For excellence you have to reach beyond yourself. It is a crucial, defining step in self-discovery, the ultimate favour you can do yourself.

When doubts assail you on the way, recite to yourself the words of Robert Browning:

> *That low man seeks a little thing to do,*
>   *Sees it and does it.*
> *This high man, with a great thing to pursue*
>   *Dies ere he knows it.*

That low man goes on, adding one to one,
    His hundred's soon hit.
This high man, aiming at a million,
    Misses an unit.
That has the world here, should he need the next?
    Let the world mind him.
This, throws himself on God, and unperplexed,
    Seeking, shall find him.

MARMALADE WARS

Norton Ashby was a figure cut from a Dickens novel, a puffed up, bombastic anachronism from a bygone age, a haughty employer careful to keep his distance from his staff, married to a devoted but invisible wife who never challenged his dominance, never gave him a moment of insecurity, of self-doubt, a chance to see himself as he really was.

At Christmas in 1949, London still black with the soot of coal fires, he surprised his staff with a luncheon at The Aldwych. The occasion was a welcome change from the drab, post-war daily routine at the advertising agency. Norton Ashby, already well into his Christmas celebration, came down from his pedestal, joking, laughing, slapping people on the back. He made a speech. The employees laughed and applauded heartily, won over by the change Christmas had wrought in this aloof man. He, in turn, saw in their laughter a sign of acceptance, an encouragement to continue. Continue he did, at some length.

At long last he concluded by saying "to the young men of this firm I have but one advice: invest what you've got in rubber and, if it comes awf, marry the girl!"

I was grateful to Norton Ashby. After all, he gave me my first job. As copy clerk at the agency I earned three pounds a week, enough to pay for food and lodging and the bus fare to work. My savings bought me a cheap seat at the cinema every two weeks.

The agency carried the account for Chivers marmalade, widely advertised as the "aristocrat of the breakfast table". The name was kept in the public eye in newspapers, magazines and in the London Underground.

The crisis came when Chivers informed us that they had been consistently outsold in the London area by Robertson's, their competitor. Shock-waves passed through the agency departments, each suddenly under scrutiny to find the flaw responsible for this appalling turn of events. Did the copywriter falter in coming up with a memorable phrase or was it the fault of the art department in failing to catch the consumer's eye? Maybe the space buyer selected the wrong media. Even the lowly copy boy was not overlooked, in case he was remiss in getting the stereo and zinco plates to the printers on time, thus helping Robertson's to reach the tape first. A state of siege prevailed at the agency. Our client was losing money and he wanted an answer. Fast.

At length, suspicion fell on advertising in the London Underground. A tough problem to investigate, with hundreds of trains criss-crossing Greater London, rolling at all hours.

It was then that I, the most junior and lowest-paid employee, was summoned before the presence of Norton Ashby.

"Lad", he began, forehead furrowed and lips pursed, searching for the right word. "You are about to embark on a momentous task, one which will inscribe your name in the annals of this old firm... it is now up to you, you alone mind, to save one of our most important accounts".

He charged me with the job of investigating the comparative advertising of Chivers and Robertson's products in the London Underground and show whether our client's advertisements were outnumbered by those of the opposition. If we could show numerical superiority, the account could be saved.

"Do you understand how important this is?"

"Yes, Sir", I replied. With the guns of the Second World War silent for some four years now, the war between brands of marmalade seemed a grave matter. As for "the annals of the firm", of their existence I had great doubt. Still I was given a job and I must do it well. The size of the task worried me a bit.

"Just how many carriages in the Underground, Sir?"

"Three thousand six hundred!", he said, as if it was just a handful.

The number was daunting. The trains were a moving target. It would take time to do the job, maybe a month, even if I went at it all day, every day.

"Report back to me in two weeks. Waste no time. You will need every minute of every day......Remember the trust I am placing in you......We are a venerable old firm, a family......

that includes you, lad, and me, Stella with her baby, Tommy with the wooden leg, and all the others......go to it!"

He handed me a pass for the London Underground System. I was free to travel, enter any station, visit depots, walk through trains parked for the night. Heady stuff.

Leaving his room I realized that I was now a soldier in the marmalade war. With three thousand six hundred carriages waiting I had about twelve useful days at my disposal. At best, it meant inspecting three hundred cars a day, a seemingly impossible figure.

The London Underground, even in 1949, was an extensive network of lines. The District, the Circle, the Central, the Piccadilly, Bakerloo, Northern and Metropolitan travelled the length and breadth of the great city. The trains hardly ever slept. I must catch them on the move.

I chose a different line each day, travelling along with each train till I had looked at every carriage, then getting off and waiting for the next. All the while, I catalogued the individual number of each car and the presence or absence of placards of the competing products. I visited every station on every line. If the same train turned up a second time, it was my loss. My new mobility was a welcome change from my desk job at first but soon it became a drag.

Worse than that was the hostility I endured from some railway crews and station staff who believed I was spying on them. They were not about to believe a soldier of the marmalade war and were vocal in their disapproval of my presence.

Daylight was not long enough to add numbers to my list. I was on the job from early morning till late at night. I visited train depots. I was shown how to open and close the doors of parked trains, turn on the lights and reverse the process at the other end.

"Just keep off the live rail!", they kept saying.

I was not paid expenses. My budget did not accommodate eating out, so I missed most meals. Still, I took pride in pulling my weight "for the family", for Mr. Ashby, for Stella with her baby, Tommy with the wooden leg and all the others. I was a team player.

By the end of twelve days I had catalogued the inspection of twelve hundred carriages, fully one-third of the rolling stock. Our advertising showed up well and had the advantage of numbers over the opposition. It took me two days to type the report. I was proud of my effort.

The Chivers account was saved.

There came no handshake, no word of approval nor one of thanks from Norton Ashby or anyone else.

I returned to my desk disappointed and resentful. I had given my best effort, worked sixteen hour days and in return I received no acknowledgement, no encouragement, no appreciation. I continued to receive my three pounds a week, a humiliating income with no margin for living.

It was time, I reasoned, to ask for an increase in pay. I telephoned Norton Ashby and asked to see him.

"What is it, lad?!", he snapped impatiently as I walked in. He was seated at his desk, coatless, a gold watch in his waistcoat pocket with an appropriate chain. Elastic armbands kept his cuffs above the wrist. He was puffing on a freshly lit cigar held in his right hand.

"Sir", I began hesitantly, though I had practised my speech a hundred times. "I did what you asked me to do.

I worked sixteen-hour days, showed my good faith to the firm, the family, as you call it and helped save the Chivers account. My salary is three pounds a week and I barely get by on it. I came to ask you to consider raising my pay ".

Even as I spoke, I noted mounting disbelief on Norton Ashby's face. His body began to quiver. In denial, his jowls swung from side to side. His mouth opened and I saw saliva trickling between his trembling lips. Fire lit up in his eyes. The impending eruption of a volcano, no less. Then, with a swift motion belying the inertia of his substantial body, he switched the cigar to his left hand, his right arm shot out in the direction of the door with speed to inspire awe.

"OUT!..!..!..!..!", he screamed.

The word was still echoing from wall to wall as I descended the stairs.

Tommy with the wooden leg was my friend. He had lost his leg in a war, I don't know which war. You could always hear his approach from afar by the mechanical click of his artificial joint. He had a cynical wisdom, fuelled by a shell crater brand of humour. Defeated he was not. I always sought him out when I was troubled. He made my big problems appear small.

He was at his desk, having lunch, as I entered. He looked in my direction in silent acknowledgement of my presence and went on examining a sausage from all sides, with singular attention.

"Widow's last chance......", he mused.

He went on with his lunch.

JELLIED EELS FOR A PRINCE

Maybe he did not know it, but the London barrow-boy who visited Cambridge Circus on Wednesday afternoons featured prominently in the cravings of those with a taste for jellied eels. I was one of them.

My friend, the old dockhand, was dying. The dust of a hard life in his wake, yet gentle, gracious, infinitely wise. Noble in all but his birth. A prince. A privilege to know.

"Can I do anything for you?", I asked.

"You can......, you can......", a faint smile, eyes closed. "I'd like to taste jellied eels again before I die." He squeezed my hand in appreciation.

It was Wednesday. I set off at once for Cambridge Circus, hoping to find the barrow boy.

I did. My purchase made, I hastened back to the hospital with my precious cargo.

Hurrying down the long line of beds to his bedside, I called out excitedly: "your jellied eels!"

He did not respond.

"Jellied eels," I whispered.

He turned.

"I open my eyes to see you," he said, "but I see only a shadow. Then I close my eyes and see a thousand stars"

His last words. Stars all around him, he died in his sleep.

An unacclaimed journey from dust to dust. No flags at half-mast. No obituary. This gentle man was not even considered a "gentleman". No tributes, no flowers.

Just jellied eels at the bedside of a silent prince.

THE QUEUE – JUMPER

Standing in line is not wasted time. It is a cleansing exercise practised in civilized societies. A ritual of self-control, a chance to withdraw into oneself, a time to think, a time to observe those who wait and those who start behind you but still end up ahead. There are always some of those.

Waiting in line is like life itself. Waiting is paying your dues to ensure that your turn will come and with it, the fulfilment of your wishes. There is no instant gratification. Life can give with a grudging hand. You pay for what you get. Almost always, you wait. Time to contemplate.

Surely, life is a ticket for a one-way trip. There is no guarantee that any station will be reached. There is no refund for incomplete travel and there is no rain-check. In today's terms it is a pretty poor deal for so singular a journey. It is not a holiday trip, either. There are hardships along the way and moments of respite are few. The best for which we hope often turns out to be second best, if that.

Hardly surprising, then, that we persuade ourselves that there is another place, a haven, or "heaven", not on the travel map, where we shall rest our travel-weary limbs, ease our pain, forget strife and enjoy peace without end. It would be too much to expect on our one-way trip, so it must follow later, if at all.

To earn admission to this peaceful destination we are taught to practise self-denial, give to the poor, to the church and to sundry charities along the way. It works like a term deposit.

The reasoning is simple. What you deny yourself makes its way to your term deposit. The more you forfeit in earthly pleasures the bigger your account. You can elect to forswear wealth, swap fame for humility and cast aside pleasures of the flesh.

Ah, but there is another way to heaven! Some jump the queue, they do not wait in line. No self-denial, no term deposit. It helps to have an Irish nurse around. I saw it happen!

I studied Medicine in the East End of London. In training to be a doctor obstetrics is the baptism of fire. Working under supervision the student learns to conduct deliveries. Most of the cases are straightforward but there is a disturbing potential for unpredictable variations. Complications may arise swiftly and may be lethal.

On my way into the delivery room one afternoon, Stella, the Irish nurse called me over. She told me this was no ordinary delivery. The child had a condition known as anencephaly, characterized by failure of the brain and spinal cord to progress from a groove into a closed tube, with resultant escape of cerebrospinal fluid.

Before the days of ultrasound the diagnosis was made by way of an uncommonly large abdominal swelling in the pregnant mother. Anencephaly is fairly rare and is incompatible with life, the baby dying shortly after birth.

In view of the rare nature of this case Stella had instructions from the consultant not to allow the umbilical cord to be cut, as he wanted the complete specimen, namely, baby, cord and placenta, preserved for the museum of the Royal College of Obstetricians and Gynaecologists.

"Does the mother know?" I asked.

"No," replied Stella curtly. "This is no time to dwell on that, she is ready to deliver. Now, you deliver the baby and wait for the placenta. You do the work, I'll do the talking. Jesus, Mary, Heaven help us!" With this, she spun on her heels, pushed open the door of the delivery room and walked swiftly to the bed. While her forward motion was rapid, she also moved sideways with each step, as if using each shoulder to hammer her heels into the ground. I followed in her wake.

"Now, my dear," she said to the patient who appeared to be in her late thirties. "Don't you worry about anything! We'll have this baby out in no time! This nice young man will do a good job. You just listen to me and do what I tell you."

Stella and I both knew that once the baby was out, delivery of the placenta may take another fifteen minutes. That would be a tough vigil for us all.

Anencephalic babies are hideous to look at. The brain is open from the top of the head to the back of the neck. The sight may fill the sturdiest of us with revulsion. For a mother, after nine months of pregnancy and hours of painful labour, seeing such a monster would be punishment indeed.

There was silence all around, except for the moans of pain from the woman lying on her side. Still, she was in control of herself.

A smell of Dettol disinfectant hung about the room. The two student nurses were at their stations, ready to hand us what we wanted. I was feeling a little nervous. Stella stood closer to the head of the bed, her arm around the woman's shoulder. Steady and comforting, a better nurse you could not find.

Another painful contraction. "Take a deep breath, hold it and p – u – s – h!!" I called and the mother responded well. The head began to show. The moment of truth was near. "P – u – u – u – s – h!" and out popped the head – a greasy, slimy apparition, quite repulsive to hold. I delivered one shoulder, then the other. The rest of the baby followed. It was a sight to remember. The face was tight and flat, with an expression as if it had just swallowed a bitter pill. The back of the head was open, revealing the brain and spinal cord. It was not a baby. It truly was a monster. We were all quiet. Stella looked at me over her shoulder. I pursed my lips to signify my horror. The child was breathing feebly. I looked at the clock. The second hand moved slowly.

The mother stirred. "Why is everyone so quiet? Stella's grip on her shoulder tightened. Her head was close to the woman's ear. "You still have work to do" she said, taking command.

"What about the baby?" the woman cried.

"When the work is done," Stella stood her ground. "We still have to get the afterbirth out".

I looked at the clock, it appeared to have stopped but it was actually working. I felt for this poor woman. It was obvious that she would be made to understand it was a stillbirth to protect her from the awful, repulsive truth of having given birth to a monster. Meanwhile we keep mum. Damn the College of Obstetricians!

Stella was marvellous. She engaged the woman in conversation, careful to steer her away from discussing the baby. I had the baby lying on a towel in a stainless steel bowl. It was still breathing, the stuff of nightmares. The student nurses went about their work in embarrassed silence. At length the placenta was delivered. I placed it in the bowl. The Royal College of Obstetricians had its trophy intact.

Stella motioned to me to take the bowl out of the room, so I headed off to the utility room to clean up, which was the student's job at the end of each case.

"Your baby will not live", I heard Stella say to the mother as I left the room.

"God's will", whispered the woman. She had other children. One less mouth to feed......

I was washing the instruments when the door flung open. Stella headed straight for me, her shoulders nailing her heels to the ground in her peculiar gait. She was earnest.

"Have you christened it yet?" she snapped.

"Never thought of it, Stella".

"Mother Mary and all the saints! You mean you are just letting his soul go to hell?" Stella was furious.

She took the bowl and turned on the cold water tap. Wetting her fingers, she made the sign of the cross on the baby's forehead.

"I christen you in the name of the Father, the Son and the Holy Ghost..." She then put the bowl aside.

I watched the ritual, touched. "What good did that do?" I asked.

"I'll tell you what good it did", replied Stella with emotion. "He dies a Christian, his soul will go to heaven which is more than I see happening to your soul when your time arrives!"

With this Stella left the room. I finished cleaning up, enveloped by the aroma of Dettol.

As I glanced at the baby he was breathing his last, his soul on the way to heaven.

His body, the umbilical cord and placenta are intact in formalin in a transparent jar on a shelf in the museum of the Royal College of Obstetricians and Gynaecologists.

THE VIGIL

The knock on my door had an urgency which made me respond at once. A distraught young woman stood outside, a neighbour in my rooming house. I saw she needed help. Without even a greeting I motioned to her to come and sit down. She sat uneasily and refused my offer of a cup of tea. I waited for her to speak. At length she broke the silence:

"I have just flushed my child down the toilet".

Her confession, not addressed to anyone in particular, asked for no comment, no consoling word. She uttered the words only to convince herself of the truth of a dreary October afternoon.

"I know you are going to be a doctor", she added, "I just could not talk to anyone else".

I was unprepared for this, my first emergency. True, I was in the early phase of my medical training. I also knew that the journey upon which I had embarked came with responsibilities, with commitment to people in need. As it happened, my first emergency arrived a little too soon. My conscience told me, ready or not, I could not walk away. I was hopelessly unready. I knew this much: the only help I could offer was common sense, observation and sympathy.

I was familiar with cadavers and test tubes. My studies had not yet reached the living.

I thought of rushing down to the public library but was gripped by anguish at the thought of the response I would receive from forbidding Miss Rumbold when I asked her for everything she had on abortion. This was the early nineteen-fifties. Abortion was illegal, I was not sure you could even utter the word within the hearing of law-abiding people, let alone Miss Rumbold.

Maybe I could go to a hospital and collar the first young doctor I saw for information? This seemed like a good idea until I remembered that this girl had had an illegal procedure and, should things go wrong, any inquiry on my part could result in a police investigation. Were I linked with an illegal procedure at the very start of my studies, it could end them there and then.

Yet I knew that women did die after illegally performed abortions. Like it or not, I was now involved. I could not back out. If only I was not so ignorant!

Here was a stranger in my room, someone with whom I had no more than a nodding acquaintance, someone who, for no other reason than trust in what I was in time to become, sought my help. I gave her asylum because, instinctively, it was the decent thing to do. Asylum meant protection, so I did accept a certain obligation. We were on our own, each in a separate way, yet brought together by the immediacy of circumstance. In those times you could not just walk up to someone and ask for advice about abortion. The rejection would be more harsh, more isolating than the state of indecision in which we now found ourselves.

"Expect to see blood", said the abortionist. At least blood declares itself, and leaves no room for doubt.

"Are you bleeding?", I asked her. She did not think so, but went to check in the only bathroom in the house, one we all shared.

She looked relieved when she returned. There was no bleeding, she said.

I asked her how the abortion was done. She did not know, but the instrument looked straight and sharp to her. Really, this was the time to see a real doctor, I told her. She shook her head, no way. What can a sharp instrument do, I pondered. It could go right through the uterus, maybe even make a hole in the bowel or some big blood vessel. There are plenty of those in the pelvis. Then, if she was bleeding slowly inside, we may not know it for hours. Bowel is another matter. When the appendix ruptures, people become ill very quickly, suffer much pain. I looked at her as she sat there, she looked comfortable. No, there was no bowel perforation.

But then, how clean were the probes used by the abortionist? Clearly, infection was a danger. I tried to explain this from a platform without authority, without conviction. At least I felt I had a grasp on first principles.

Surely, if hemorrhage from the uterus were to occur, it would happen early, I reasoned. If the probe cut through a big blood vessel, blood loss would soon lead to shock. She was not pale or dizzy, so that was out. As for infection, that would take time to show. I did not know how long.

God knows what other problems awaited this girl following the loss of maybe the only child she was ever to have? Questions, questions, no answers. Darkness, unrelieved by friendly light.

We ate the meal I prepared without even noticing its contents. The cup of tea was comforting, as was the silence in the room. There are times when, try as we may to isolate ourselves, we still depend on the presence of our fellow-man, when giving our time to someone in need is not a choice but a spontaneous response, a sign that in some obscure way we are all linked with one another.

At length the silence had done its work and she began to talk. I listened without interruption. What unfolded was a tale of hopelessness that I would not have perceived behind the friendly nod she always gave me on the stairs.

She was an immigrant, flotsam in the wake of the war not long over. She lived alone, had no family or friends, worked in a one-man office to which she commuted on an old bicycle she bought in the flea-market. Her employer had made advances to her which she refused. When she told him she was pregnant he made clear that she would have to find other employment once her pregnancy began to show. She lived from hand to mouth and her existence depended on everything staying normal at all times.. She had no safety margin, no money to spend on pleasure, no prospects. Her only emotional contact with a man had occurred in a concentration camp during the war. They kissed across a barbed wire fence, and that was that.

Along came Ashley, tired of the countess who had been his mistress. He was debonair, had a good war record and he was intent on seduction. She could not resist him. They had a brief, intense affair during which Ashley was a frequent visitor to her room. The visits stopped when she missed her first period. Her periods had never been irregular before, still she kept on hoping. When the second period failed to arrive she went to see a doctor. She was desperate.

The doctor shook his head. "Sorry, I cannot help you. You know it is illegal. Next time be more careful!" Next time…………………

She heard of an abortionist who worked above a fish and chips shop. Rumour had it that the instruments were sterilized in hot fat. It was a shadowy world, operating outside the law. Contacts were made by word of mouth, the weak link in such an operation. One errant word, a trust misplaced and the law brings the business to an end.

The abortionist had no sympathy. "You got yourself into this mess and now you are asking me to risk going to jail! Of course it will cost you. You pay up front, thirty pounds."

Thirty pounds represented her income for a month.

"I have no time for small talk. I don't want people coming and going, so let's get this over with. Put the money on the table before you remove your clothes And, for God's sake, keep quiet! I want no screaming. I have a business to run."

The girl did as she was told. Intimidating and unpleasant as the situation was, it gave her a chance to put the problem behind her. The abortionist was surprisingly deft, she had done it before. She went about the job in silence.

"Forget my face, forget the address. You are on your own now. You will see blood. You will see the creature that was your child, there is nothing abnormal about that. Watch out for belly-ache, watch out for fever. Go and see a doctor, if you must, but on no account give me away."

The door closed behind her before she knew it. A smell of fried fish wafted down the street as she walked away.

It was already dark when she stopped talking, apparently relieved by recounting her story, words to explain her intrusion into my afternoon. She smiled the wry smile of one who knew the game was lost. Still, thus far, nothing untoward had happened. In fact, she seemed better, her acute distress had drained away.

Yet, the vigil was not over. I suggested that she return to her room, try to sleep. I would come and take her temperature at two and again at six. I set my alarm.

When she had gone I went to the public library to see what I could find on the subject of abortion. Miss Rumbold sat at her desk, her head motionless, her eyes following every movement in the room through wire-rimmed glasses. She had intimate knowledge of the books on the stacks and could read your interests and intentions by the aisle into which you chose to turn. The section about health and hygiene was poorly stocked. I looked at the titles with feigned disinterest and, in order to confuse Miss Rumbold, I took several books from the shelf once I spotted a title which looked promising. I even took a volume from a nearby shelf on the subject of oriental carpets, to serve as a decoy. Hardly likely to fool Miss Rumbold but I was engaged in clandestine activity and found her authority daunting.

At a table out of her sight I very soon located the section which contained the information I wanted. I read quickly, turning the pages impatiently, breathless with anxiety.

Someone had beaten me to it! The pages which, surely, would reveal the information I sought, were missing, torn from the book, not very neatly at that. So that was that!

I gathered up the books and returned them to the bin, Miss Rumbold's eyes following me as I did so. I gave her a respectful nod on the way out, which she acknowledged with the merest, unsmiling nod of the head.

The girl was in a deep sleep when I saw her at two o'clock. She had no fever. When I returned at six she was already up and she gave me breakfast of toast and coffee. She was feeling fine but I took her temperature anyway. It was normal.

"Please see a doctor today", I pleaded before I left for my classes.

"Maybe I will" she replied.

On the long train journey to my class I realized that I had passed a new landmark in my life. Although I felt sleepy and tired I had a new sense of worth, a better understanding of where I was heading in my studies. The trust this total stranger had placed in me just because she knew where life would take me was at once bountiful and humbling. You do not become a doctor by merely passing examinations. You do it also through tiresome vigils on dreary October afternoons.

In the weeks that followed she continued to feel fine. On the occasions when we met on the stairs she gave me a knowing smile. We had a secret which would remain a secret, always.

Even months later, whenever she passed me in the street on her bicycle, she would pull over for a friendly word or two. She still remembered the afternoon in October.

One day, later still, she rode by me, called my name and waved. This time she did not stop to talk.

The vigil had ended.

OFFICIAL SECRETS

The people around us, those we love, or just know, or maybe those we simply cannot avoid, fit together like pieces of a jigsaw. With the passage of time, of the jigsaw we knew only fragments remain. Memorable or not, deserving or otherwise, some occupy our recollections like squatters who have no intention of leaving.

KLARI

Klari was kind-hearted. Her table, generously laden with food was always surrounded by a lively crowd. The fine, sprawling apartment in Portland Street allowed for the guests to break up into groups. Hospitality was beyond reproach. Of all the invitations I received as a boy in his early twenties hers are the ones I most fondly remember.

I must confess that at the age of twenty-one I was still burdened by a pathetic innocence, born of the academic challenges I had set for myself which focused my entire attention on passing the next set of examinations, never more than weeks away. While I saw life all around me, I failed to observe the nuances which for others were a rich source of gossip, entertainment and insight into the private lives of others.

It was a twenty-year old, beat-up Bedford van which awakened me to the fact that my narrow field of interest had blinded me to the drama, subterfuge and intrigue going on around me. Heading for the sculptured, ornate front door of Klari's building one day, there was not a customary Bentley, Daimler or Aston-Martin in sight. Instead, incongruously, there sat this mongrel among automobiles, this well-worn, dented jalopy of a dirty old van which paraded its vintage and humble pedigree in defiance of the opulent face of the neighbourhood.

Klari answered the doorbell with a hug. Almost protectively, albeit with good humour, I called her attention to the mutt parked by the front door of her elegant building. Klari paled and bit her lip, seemingly caught off guard. She regained her composure and conspiratorially informed me that she had taken driving lessons and the van was in fact her car. Her first car! "Don't mention it to my husband", she said. I squeezed her hand in congratulation and changed the subject. Within the apartment conversation was already spirited. Soon, with glass in hand, I was seated on a sofa with some of Klari's friends. It did not take long before talk shifted to the Bedford van outside. At once the volume dropped. Instead of merriment or criticism I detected envy in the whispered exchanges. That they continued to talk in my presence indicated that I occupied no place of importance in their world.

"Klari has a lover", one of them said.

"High time!", observed another.

"With a man like her husband I would have one too".

Like a non-combatant in no-man's land I elected to lie low, stay mum, studying the ice in my tumbler. The revelations flowed.

"The man lives in Kensington. Buses and trains are so complicated! Taxis are expensive. So Klari decided to take driving lessons and have her own transport. Now they can meet wherever they choose".

"But why such a dilapidated car?" asked one in anguish. "You can see it a mile away! Why not a Morris, a Hillman,......... an Austin?".

"Klari's husband is not generous", remarked one in resignation. "Just like my husband. Maybe I should find a lover......".

"Klari saw an ad in the Evening Standard", chipped in the third.

"1932 Bedford van. Fifty pounds. Good runner. Body a shocker but what do you expect for fifty pounds?"

"Dinner's ready!", called Klari.

LILY

In this strange circle the one person to whom I could talk about my problems was Lily. Lily was fortyish, divorced, living alone. Always ready to listen, she would zestfully light up a cigarette and appear to give full attention to the confession she was about to hear. She seemed accustomed to hearing confessions. In a way it was like talking to a statue, a chance to unload what was on my mind, without expecting any response or judgment. But then, did Francis Bacon not say that it was better to unburden oneself to a statue than to dwell on problems in solitude! Yes, I liked Lily a lot. Of all the people around me she was the one with a heart. Or human feelings, at least.

Other people liked Lily, too. Especially the men. She carried her femininity well and, as I came to learn, most of the men in this circle had found in her a good listener and indeed a woman who could dispense her sympathy in more generous ways. Of course, discretion prevailed and any hint of a tryst was tacitly brushed aside by everyone's good manners. The men, if jealous, did not seek to find fault with her because they knew that, given the right circumstances, they could turn to her bowl of compassion for comfort. An outsider in this drama, hardly twenty years old, with no one to talk to, no shoulder to cry on, Lily was my favourite character in the play.

I was an ill-fitting piece in this jigsaw of oddly-assorted individuals. Twenty or thirty years younger than any member of the group, I played no role of interest to anyone. I came into their midst by happenstance. At dinners and cocktail parties I was expected to dance with the women when the band struck up. On such occasions I felt an obligation to make small talk around the floor which in hindsight was good training, albeit tedious at the time.

Stephen found it irksome that I had no girlfriend. In his book a young man had to have his "baptism of fire" and I was clearly behind schedule. His picture of me as a young man about town was oblivious of the simple fact that I had neither time nor money for new relationships. When once I told him I had taken a fellow-student to the cinema he responded with icy silence. He was not interested.

Lily invited me to her apartment for dinner one night and over delicious food and a fine claret told me of Stephen's impatience with my dogged virginity. I told her of my circumstances which simply could not accommodate distractions, departures from my Spartan routine. I was on a tightrope and had to pay full attention, lest I fall.

Quiet music filled the room. The rest of the world was far removed from the little apartment. Lily lit another cigarette and began to talk about everyone's need for physical love. Through clouds of smoke she told me of the fulfilment it can bring. As she reclined in her armchair I could understand why all those men were attracted to her. Her voice was sultry and wistful. Her eyes did not leave mine. She spoke slowly, leaving long silences to allow her message time to sink in. And then she told me of one particular aspect of lovemaking that gave her the most pleasure. "Oh, I love it!", she said breathlessly………"I love it best of all!",

Although I left the apartment without accepting her obvious invitation, I was grateful to Lily for the overture she had made. For the first time in my life a woman talked to me openly about physical love. She allowed me a glimpse of herself, a risky, generous gesture, one that did not diminish her in my view. There was nothing in it for her, she reached out a hand to a young man as only an understanding older woman can. In a way I felt I had let her down but I hoped she would understand. She knew as well as I did that if I had weakened and given in to the compelling dictates of the moment it would have opened a door to circumstances I could not control, to occasional fleeting, clandestine pleasures bought at the price of regret and disillusion, conflict and guilt, diverting my attention from the routine I knew, one that if I only stayed the course, could guarantee my survival.

Later, Lily found security in marriage to a delightful older man. A year or two later still I was disappointed though not surprised seeing a handsome young man leaving her apartment one afternoon, Perhaps yet another confession, another gift of her compassion.

She died unexpectedly, under mysterious circumstances.

Rest in peace, Lily. I love you.

CHARLIE

Charlie was a bounder, a cad who devoted his entire time to the pursuit of his own happiness. He cut a wide swath, letting everything fall in his path. That path, of course, was littered with women.

His wartime service had prepared him for the life he was to lead after the guns fell silent. In or out of uniform, his work was hush-hush, as was his life. By inference no one ventured to ask where he was going and why. He had long absences, supposedly under the umbrella of the Official Secrets Act, although I later came to recognize that the guise of official secrecy served to sweep some personal secrets under the rug.

Be that as it may, Charlie was a most presentable man. He had a fine profile rendered more memorable by an aquiline nose. His hair was brown and straight, parted exactly right to make the best possible impression from whichever side you chose to look at him. He was of average height, diffident of manner, soft spoken, his lips hardly parting as he talked.

Perhaps a relic from his wartime role, whatever it was, that he never talked to anyone for long, as if to leave as few traces of himself as was possible in the circumstances.

Charlie, fluent in German and French, was married to a fine-looking German woman he met while engaged in secret work in the aftermath of the war. She left so little of herself to recall that she could easily have been a secret agent herself. Perhaps she was. In any case, I cannot remember one idea, one sentence, indeed one word emanating from her lips. To all appearances they were well-suited and they attended all the parties together, although I noticed that on such occasions they always drifted to opposite ends of the room.

It was known to all but mentioned by none that Charlie had a heavy, ongoing relationship with The Baroness, part of the circle, a woman living alone, flotsam from the aristocracy of yesteryear whose redundant title in the post-war twilight was no longer a magnet for aspiring social climbers. Whether she had discarded the Baron or he discarded her, was not public knowledge. She had an affected air and generally looked down on others of less than aristocratic rank. At parties she would position herself with Charlie in her line of sight and fire off lovelorn, long distance glances in his direction.

Gossip flowed like a river in flood at such gatherings, most of those present smoking, as was the custom of the day. Strangely, Charlie never became a subject of wayward talk, perhaps out of patriotic deference towards his war record. Yet, just what that record was no one knew. So I was more than surprised one afternoon when I found him hurriedly climbing the stairs at my modest lodgings. He did not stop to exchange conversation. A man in a hurry, he went on his way. Perhaps more secret work......... In my rooming house?

As it turned out, it was secret. In his peregrinations Charlie had encountered a young spinster who made a convenient diversion from the formal, more ritualistic encounters with the Baroness. His visits for "afternoon tea" became regular. Regular, that is, until pregnancy put a stop to afternoon tea. Thereafter Charlie was not seen again on the humble premises.

Charlie was always courteous towards me and perhaps inwardly grateful that I remained discreet about his assignation. He was popular with the women and they felt secure with him, knowing that whatever indiscretions they might perchance commit in his company would be as good as covered by the Official Secrets Act.

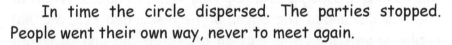

In time the circle dispersed. The parties stopped. People went their own way, never to meet again.

As for Charlie, he just vanished from the scene.

CENTRE OF THE WORLD

My late old friend and colleague Sam Huang told me that the name "Asia" originated from what I can only phonetically describe "AWE JOW", old Chinese for "the centre of the world". You cannot blame the Chinese for that designation. Thousands of years ago, when all travel was made on foot, horseback or by perilous ventures to sea in flimsy boats, their concept was certainly justified. Even now, in the twenty-first century, for many of us home is the centre of our world.

    I spent the nineteen-fifties in England, most of that time in London. It was the centre of my world for sure, not least because my constrained finances prevented travel beyond the reaches of the London Underground, for which I mercifully possessed a season ticket.

    As it happened, my attention at that time was focused on my studies and I paid little heed to what went on beyond the confines of that great city. Besides, to a Londoner, as I regarded myself, everything that mattered seemed to take place there and if I did not go out to visit the rest of the world, the rest of the world came to my doorstep, or so it seemed from what I read in the newspapers or heard on the radio. Yes, I lived in "the centre of the world".

That being so, it was no surprise seeing the philosopher Bertrand Russell sitting at a nearby table in a restaurant, the nabob Nubar Gulbenkian enthroned in his custom-made London taxi in the next lane at the traffic light, the breeze blowing the white hair of the actor Spencer Tracy in a convertible alongside the bus in which I was riding, or coming face to face with the actor Alec Guinness at the pedestrian gate into Green Park, allowing him to pass. Little did they know it, but they shared London with me for those moments and at the end of the day they gave me something to remember. You had to be at the centre of the world to experience such encounters. Yet, far beyond such "sightings" of men of renown, it was the crossing of paths with a war hero which preyed on my memory and fuelled my doubt for decades.

Though memories of it were slowly fading, the Second World War was still very much in the rear view mirror of life. Rationing for food and clothing had hardly ended and scars from aerial bombing still pockmarked the face of London.

Electricity was creeping into areas that had hitherto been lit by gas, an eerie, greenish glow, not fondly recalled. Chilblains in the fingers were still an aggravation in houses poorly heated by coal fires and gas stoves and the single pane windows did little to keep the chill out. The devastating London smog of 1952 enveloped the city in a green shroud, limiting visibility in daylight to some fifty yards, much less at night. The choking green air invaded every conceivable breathing space, the streets, the trains and your very room and caused some three thousand deaths among those whose lungs were already doomed. I recall going to the latest American movie, supposedly in Technicolor. The fog translated it into a turbid, green spectacle. I crossed from Waterloo Railway Station to the Westminster underground late one evening, navigating by the touch of my hand along the London County Council building and the length of Westminster Bridge. In truth, it was exciting. You had to know your London.

What made this period remarkable was the "never-say-die" spirit of London. Amid the soot, the smog, the food shortages and chilblains there was an unmistakeable zest for life. Not taking hardships lying down, the Brits got up early in the count, put their best foot forward and resolved to change gear towards a better tomorrow. You could tell it by the tone of their voices, the spring in their pace, the shine in their eyes. It was nothing short of inspiring. It was infectious, so early in my studies it could not have come to me at a better time.

Having seen the war at close quarters, I was still trying to make sense of what I had experienced. I read every account of the war I could lay hands on. The soldiers, the airmen, the Resistance and the secret agents parachuted into Europe were my heroes. I owed them a debt for just being where I was, for the opportunity, once so remote, of studying toward a degree in Medicine. They became my role models: I translated their fight against the odds into my quest to become a doctor.

So it happened that one Saturday afternoon, while living in the Earls Court area of London, I spent time reading about the remarkable war of Group Captain Leonard Cheshire, winner of the Victoria Cross, Britain's highest decoration for bravery. Cheshire was an exceptionally skilled pilot whose courage under fire set him apart. I learned that he flew a de Havilland Mosquito as a Pathfinder for Bomber Command, flying at low altitude to the target and while being fired upon by every enemy anti-aircraft gun, marking the targets with coloured flares to guide the bomber stream that came behind him. According to reports he stayed at low level while the bombers above discharged their bomb load and radioed information to those that followed if corrections were necessary. Truly an inspiration.

Immersed in my book, I looked at the clock, only to discover that the hour was running late and I had to go out to buy food for the weekend. As I rounded the corner into Barkston Gardens, a block from the Earls Court Road, I came upon an old school bus parked there, with a slender man standing beside it. He told me that inside the bus there was a replica of the Shroud of Turin, a relic which many believe bears an imprint of Christ`s face which it covered after the Crucifixion. I accepted his invitation and entered the bus with him. He quietly pointed out features of the exhibit. I thanked him and left.

As I walked away, I had the feeling that I had seen him before. Slowly it dawned on me that his was the face of the pilot in the book I had just been reading! He must be Leonard Cheshire, I reasoned! The chances of my being right were minimal. I did my shopping and headed home. By the time I turned the corner into Barkston Gardens he was gone.

For decades thereafter I thought I must have met Cheshire himself. "Must have", but had no proof. Did I meet him or did I not? What would a war hero be doing travelling in an old school bus with the copy of a relic which itself may or may not be genuine? Should I have rushed back to ask him if he really was who I thought he was? You just don't go up to someone and ask such a question. An outstanding figure of history he may have been, but he was gone and I stayed alone with my doubt.

Sixty years later I visited the Aviation Museum at the Edmonton, Alberta, City Airport, where there is a beautifully preserved de Havilland Mosquito on display. While I stood admiring it from every angle, a small group of volunteers, advanced in years but all younger than myself, joined us and we talked airplanes. "Hangar flying" is a habit of pilots and we delved deeply into the history of the airplane. As it happened, being younger than myself, they knew nothing of the famous Pathfinder Squadron, nor even about Leonard Cheshire himself. They listened intently as I recounted what I knew of the air war, the role of this airplane and, in particular, the part of Cheshire in that setting. I am sure that they now impart to new visitors the story they had just heard. Eva stood quietly in the background, letting "the little boys" enjoy their conversation. We shook hands all around and left with a good feeling.

Still the nagging doubt remained. Was it really Cheshire I met in the bus at Earls court? On reaching home I turned on the computer and there it was: after the war Cheshire, who, incidentally, had witnessed the atomic bombing of Nagasaki in an accompanying plane, devoted his life to good causes. Among other things, ".He travelled Britain in the Fifties with a replica of the Shroud of Turin", stated the text.

It took fifty years but I had my answer: I did meet my hero in an old school bus at Earls Court that grey Saturday afternoon.

London: centre of the world............ my world........

BERNIE'S WAR

At five foot five, maybe six, Bernie Stutter F.R.C.S. stood tall among surgeons. The excellence of his surgical skills was secured by sound judgement. He was able to work seemingly endless hours with never a harsh word, remaining approachable and congenial all the while. He was a pleasure to work with, a compassionate doctor, a patient teacher, with kind authority glinting from his spectacles as he looked up at you. The surgeon to have around if ever you needed an operation. Unlike many surgeons of his day, he carried a humility which did nothing to diminish the trust he instinctively generated.

You could say in jest that his small stature, making him a smaller target, might have served him well in the war had he been an infantryman. Alas, in his theatre of war it helped him not at all. He spent much of his war as Surgeon Lieutenant on Royal Navy ships escorting convoys to and from Murmansk, in arctic waters. Carrying supplies to the Russians along this route was one of the most dangerous undertakings in the Second World War. As they sailed north from Scotland they were prey to German submarines lying in wait. The convoys had to move at the pace of the slowest ship in order to maintain cohesion, thus increasing the hazard of contact with the enemy. As they turned east they came within reach of the Luftwaffe stationed in northern Norway. As if these perils were not enough, the Arctic Ocean mounted its own attack in the form of bitter weather, heavy seas and icy winds. The spray froze on the superstructure and ships were known to become top heavy and founder in the process. In those waters survival was out of the question, not only because of the extreme cold, which kills in minutes, but because the other ships had to keep going. There was no provision for rescue. A man's fate was tied to that of his ship.

Lesser men would carry this experience as a wound for life. Not so Bernie Stutter. A happy survivor, he had overcome the dangers, dodged the threats from the sky, sidestepped the onslaught of arctic storms and the sly menace of the U-boats and had every reason to consider himself bigger than all the challenges thrown in his path. He did not dwell on the hardships and had no recurring nightmares. Rather, he fondly recalledamusing incidents of his time as a ship's surgeon.

Reaching home port after a particularly perilous trip to Murmansk and back he was packing his things in anticipation of his forthcoming shore leave when he was summoned to see the captain.

"Stutter", said the captain, looking haggard after the wearisome trip, "awfully sorry to tell you this, but you'd better get your things, you are sailing with the outgoing convoy this evening".

Crestfallen, barely out of the jaws of death, Bernie countered defiantly: "But Sir, may I ask why?"

"The medical officer on another ship has appendicitis. You are to replace him. So sorry......"

At once Bernie went over to the ship in question, asking for the ship's surgeon. "Just follow the sound of the party, Sir", said a sailor, nodding in the direction of the revelry. Descending into the bowels of the ship Bernie found him all right, celebrating his acquittal, hale and hearty, beer in hand, in the company of his fellow officers. Hardly a case of appendicitis, the man was swinging the lead, just trying to get out of an unpleasant assignment. Bernie confronted the man, examined him and found no sign of appendicitis. Forthwith he returned to his ship to see the captain.

Pulling himself up to his full five foot five, maybe six, he saluted and announced with emphasis: "I refuse to sail on the outgoing convoy, Sir! "

The captain took the pipe out of his mouth and looked at him with mounting disbelief.

"But Stutter, this is mutiny!"

Bernie stood his ground. Resolve is not measured by a man's height. He put the facts to the captain who, though still somewhat shaken, nevertheless understood the gravity of the order he had previously given and promptly informed the captain of the outgoing vessel that his ship's surgeon was fit to sail.

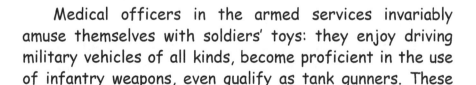

Medical officers in the armed services invariably amuse themselves with soldiers' toys: they enjoy driving military vehicles of all kinds, become proficient in the use of infantry weapons, even qualify as tank gunners. These diversions make a pleasant change from the daily round of routine examinations of essentially fit men. Bernie Stutter was no exception.

With his ship tied up in Tobruk he disported himself driving landing craft, laden with casualties for transfer to the hospital ship, back and forth across the harbour. No submarines here, no Luftwaffe attacks, no icy arctic storms. Just a bit of afternoon fun with military equipment while soaking up the sun.

On his umpteenth trip across the harbour he saw an ominous shadow bearing down on him. When he turned he saw the battleship *HMS Prince of Wales* heading straight for him, perilously close. Semaphore flags fluttered more in urgency than in celebration of an otherwise sunny afternoon. A collision seemed inevitable until at the last moment the battleship altered course sufficiently to ensure safety of the landing craft. Now it should be known that a battleship has right of way and will not alter course for any water craft in its way. Never!

Upon completion of the crossing Bernie was immediately summoned to the presence of the shore commander.

"What the devil are you up to Stutter?! The Admiral is furious! You caused him to do what he would never countenance doing! You made his battleship give way to your landing craft! "

Bernie winced at the onslaught. He did not mean to spoil the Admiral's day. Hesitantly he asked the shore commander if there were to be consequences.

"Consequences, Stutter?! There should damn well be consequences! You should be hung, drawn and quartered,…………but, as it happens, you are off the hook.

Did you not see the signals flying back and forth while you were busy with your toy?"

Semaphore was not one of Bernie's strong points, so he just let the man continue.

"While you were playing sailor the Admiral sent a signal at 1705, requesting the name of the officer in charge of Landing Craft 065. I responded by telling him it was Surgeon Lieutenant Stutter."

Bernie's knees were ready to buckle at this unwelcome disclosure. The man went on:

"At 1706 the Admiral sent a further message:
............*disregard signal of 1705*"

When in 1941 the entire Royal Navy was searching the North Atlantic for the surface raider German battleship *Bismarck*, then the largest, deadliest vessel of its kind, Bernie served as ship's surgeon on the cruiser *HMS Suffolk*. The patrol was successful. *HMS Suffolk* played a significant role in spotting the enemy. The *Bismarck* was hunted down and after a fierce battle, sent to the bottom of the sea. Upon return to port the ship's crew was informed that the King was to pay a personal visit to congratulate everyone on board.

Bernie was uneasy about the prospect of shaking the royal hand. King George's problem with stuttering was widely known and Bernie was fearful that when he was introduced as Surgeon Lieutenant Stutter the very sound of his name might precipitate an embarrassing paroxysm of the

Monarch's handicap. Still, he was confident that the captain, well aware of the problem, would discreetly bypass his real name and introduce him as Collins, or Watson.

The King was piped aboard and the weary sailors were introduced, one by one. The King was smiling and even had a little chat with some of Bernie's shipmates as he approached along the line. Bernie's apprehension evaporated and he was now eager to come face to face with his king. After all, they had fought "for King and country" and the visit to the ship was a pretty decent gesture of appreciation for what they had done. The Atlantic gales, the biting wind and the perils from sea and air seemed a fair price to pay for a personal encounter with one of history's figures.

At length the King stood before him and as he offered his hand to shake Bernie's the captain proudly announced: "And this is Surgeon Lieutenant Stutter, Sir!"

His Majesty, still smiling, said: "P-p-p-leased to m-meet you, Stutter......"

At that moment Bernie felt like a traitor, a treacherous, disloyal assassin, a despicable betrayer whose entire life would henceforth become a helpless, impotent negation of the moment that had just passed. He had no further wish to live. Close by, the railing on deck appeared as the inviting barrier between life and death if only he would break ranks now and hurl himself over it into the sea.

In the event, he did not leap overboard. Lucky for me, else I could not have written this story.

LIBERTÉ, ÉGALITÉ, FRATERNITÉ

Did those heads roll for nothing?

If I alleged that I have read every book about the French Resistance in World War II you could accuse me of mild exaggeration. Yet I would be pretty close to the truth. The resistance fighters were my heroes and I knew their stories. Many perished, some survived. After capture, many were taken to 84, Avenue Foch in Paris for interrogation and torture at Gestapo Headquarters.

Returning to Paris years later, a visit, should I say, pilgrimage, to 84, Avenue Foch was high on my agenda. I was back again in the land of the sidewalk cafes, where every whiff of *Gauloises (bleu)* and every strain of accordion music drifting in the air spoke but three words: *Liberté, Égalité, Fraternité*.

I set off along the imposing Avenue Foch toward my destination. I hummed a *chanson* just to set the mood. In the centre lanes the traffic was lively. A narrow service road ran on both sides to give access to the buildings. These were not ordinary houses but elegant, imposing structures, keeping their distance behind forbidding fences which served the purpose well. A heavy aura of wealth hung about the avenue from one end to the other.

My pace quickened as I approached No. 84. At last there I stood, facing the building so familiar to me from the books I had read. The names of the men and women who were brought here flashed through my mind and I knew there had to be many others whose stories were never written. I thought about what they did and why. I stood in reverence, as if at a shrine. It was a moving experience, a destination reached.

I spied a narrow roadway at the side of the building with a wide open gate. A sign read "Private Street." I entered and found myself in a round, bottle-shaped place, imposing buildings side by side, with the roadway representing the neck of the bottle. Just as I had hoped, I was now looking at the back of 84, Avenue Foch. More images, more recollections. A pilgrimage, for sure.

A minute or two later a figure appeared at the neck of the bottle. He stood there looking at me with an expression of quiet impatience. I was soon finished with my private ritual and slowly began to walk toward the only way out. He barred my way.

"Monsieur, this is a private street," he said. He was about 65, dressed in a grey suit, unsmiling.

"Please excuse me," I replied in my broken French, "but I just had to come here and look at this building. You see, I remember the war well and I know that heroes of the Resistance were tortured here. I wanted to see the building and remember them."

"Monsieur, this is a private street," he repeated without moving a muscle in his face or recognizing some merit in my motive, even if it was about heroes.

I had assumed at the outset that a man of his vintage would drop the official mask and join me in a moment of camaraderie in remembering fallen heroes, even if he then had to make it clear that he was there to do a job. I was wrong.

"A private street," he repeated and stood back to let me pass. At that moment I felt disillusioned and hurt. It seemed that all that talk about *Liberté* was just a sham.

I turned back into Avenue Foch in the direction of the Étoile. Almost at once I saw the sharp fangs of a vicious dog snapping at my knees from inside the fence in a frenzy of impotent anger. The wrought iron fence was covered over by a wrought iron sheet, painted black. An unfriendly barrier, designed to prevent any view of the courtyard within, save for a foot-wide strip at knee level which allowed the dog inside to show his stuff every time a pair of knees happened by. I had never had an encounter with a vicious animal like that. I knew, too, that he was just the envoy of people inside who approved such treatment of passers-by.

"*Fraternité*….?" I asked myself and chuckled in disgust with the dog's last salvoes receding behind me. Well, I thought, at least everybody is equal. That must count for something.

Up ahead in the next block stood a young girl dressed in a tight leather jacket and a very short leather skirt. She seemed to be waiting for someone. Just then a Mercedes-Benz peeled off from the avenue into the access road. The girl moved quickly into the roadway and opened her leather jacket. The Mercedes slowed down, but did not stop. A shiny Citroën followed and the jacket was flung open again as she stood in a provocative pose. The Citroën maintained speed and rolled by. The girl returned to the sidewalk. In no time a 2 CV cabriolet, not a vehicle for the affluent, put-putted along. The girl stayed on the sidewalk and looked the other way, her jacket firmly closed over her chest.

My disillusion was complete. There was no *Égalité* here, just as *Fraternité* was a joke and *Liberté* a sham. The French Revolution had failed. All those heads rolled for nothing!

Sullen pedestrians strode by in the zigzag manner peculiar to Parisians who are anxious to avoid stepping on dog-droppings. Should I tell them about my discovery? I thought it over and decided against it. I felt smug about leaving them in ignorance.

As I got back to my hotel a whiff of *Gauloises* drifted down the street and nearby an unseen accordion was playing.

No use fighting it!

THE ART OF WARFARE

Is it a sport, a religion, a "cause", maybe a search for identity, a means of self-expression for couch potatoes or an affirmation of loyalty and faith by diehard fans who pack the stands in blizzard conditions to support their team?

Canadian football is all of the above. It is an integral part of the calendar. During the short playing season the teams in the Canadian Football League define the identity of the cities and provinces they represent. Never mind the fact that many, if not most, of the players come from the United States. For the duration of the season they are Canadians, their helmets bearing the maple leaf, the proud symbol of Canada. Fervor becomes frenzy as the teams progress toward the ultimate goal, winning the Grey Cup.

"The Garrison Commander will see you presently", said the Major as he emerged from the room next door. This same Major telephoned me the day before and in an imperious tone informed me that my article in the Sarcee Village News had offended the local army brass. On joining my regiment, unwittingly I became editor, scriptwriter, publisher and typist of this publication which served to keep the population of the married quarters current with I don't know what.

Each month I spent time putting the new issue together, trying to inform and to amuse the reader. When I deemed it right I wrote about problems experienced by the residents of the area, in this way becoming a spokesman for those who had no standing to speak up for themselves. Hence my summons.

"Smitty" Gardner was a Calgary surgeon of renown, a mountain climber and a free spirit who became a mentor and a friend. It was his influence which made me choose surgery as a career. He had a house guest from England and he felt that the proper way to introduce the man to things Canadian was to take him to the Calgary Stampeders football game to be played that afternoon. As luck would have it, he was called to operate on an emergency just as they were preparing to leave. His wife Laura then took the visitor to the game while Smitty drove to the hospital.

While waiting to see the Garrison Commander, a colonel, the Major went on with his work. I did not like the man because he talked down to me. Maybe rank was important to him. Our mutual dislike was only too evident. While shuffling paper on his desk, he remarked, without even turning towards me:

"Captain Rety, you are a bit of a shit-disturber" I did not like his version of me, so I retorted:

"Sir, you cannot disturb it unless it is there to disturb". This was neither erudite nor courteous on my part but I had to retaliate.

He remained silent.

When the bell rang, he briefly vanished through the door, then announced:

"The Commander will see you now".

I stopped at the door and gave the Colonel a smart salute. He was seated at his desk which was clear, save for my newspaper with the offending page uppermost.

"Good afternoon, Sir! How did you like the last issue?" Always shoot first, was my maxim.

"What?..... what?....come and sit down. Major, please close the door".

The Garrison Commander was a kindly man, nearing the end of his Army career. A man of medium build, and kindly disposition, wearing thick glasses, his manner avuncular. I liked him instantly. It was only later that I learned that he used to listen to my lectures to the soldiers, out of sight, around the door of the lecture room. He dismissed the newspaper issue and asked me how I liked living in Canada. Was my accommodation comfortable? Did I like the work? Far from a disciplinary procedure the visit became a social encounter.

And then he asked me if I played chess. While not a grand master, yes, I enjoyed the game at my own level. He then proceeded to explain to me that Canadian football was really a game of chess. As in the latter, there were numerous pre-planned moves called "plays", each designed to deceive the opposition and ensure forward progress of the ball until it crossed the opponent's baseline. In the huddle the attacking team finalizes the move about to be played, assigns roles to individual players, to be acted out when the whistle blows.

"Just like warfare", was his assessment. He rang the bell. "Would you bring me my chess set", he called out when the Major opened the door. The Major's response took maybe two seconds, it was not the prompt "Yes, Sir!"

The Colonel then went on to expound the fascinating parallel between the game of chess and military battles. He was in his element and it pleased me to see him so enthused. Of course, he was an ardent fan of the Calgary Stampeders football team.

We set up the pieces and he illustrated some key opening manoeuvres. Every move counts, he said. "Gain control of the centre, mobilize knights before bishops and open up lanes for your big guns, the rooks!"

He rang the bell again.

"Would you bring us some tea!" he asked the ashen-faced Major.

I gained a new appreciation of this kindly man. As I left, he lent me a book on chess by the great Master Irving Chernev. I also learned to look on the game of Canadian football with a new respect.

Laura and her guest were just arriving home from the football game as Smitty drove up the driveway.

"Well, how did you like our game of Canadian football?" asked Smitty with an expectant smile.

"Oh, Smitty, what an experience! The atmosphere, the action, the pace, all those big men piling on top of each other! Amazing!"

Smitty was pleased to see his guest's enjoyment of a new experience.

"Who won?", he asked his friend.

"Now that is a difficult thing to say, Smitty! But if the conversation behind us is anything to go by, the bastards beat the sonofabitches by sweet fuck all!"

MIRROR WITH NO IMAGE

Although I tried my best in varied pursuits as a boy, I lacked faith in my ability to do things well. I received plenty of encouragement at home, yet I felt, even at an early age, that home-grown approval was devalued by the loving bias of my family.

It is natural to want to know how far you have walked, how fast you have run, where you stood in the scheme of things. Your own face is a stranger to you until you view it in a mirror. Independent approval and honest criticism serve like a reflection in which you can see yourself. Seeing yourself for what you are is a starting point.

An early casual comment taught me that a seemingly trivial remark may have profound effect. At age eleven I was playing soccer for my team when I found myself before an open goal yet I failed to score because I could not kick the ball with my left foot. We lost. After the game my best friend's father shook his head and said "Miki, of all people I expected you to be able to use your left foot!"

I could not sleep that night, the remark echoing in my brain. For the first time, someone made me come face to face with one of my failures. I spent the next afternoon kicking a soccer ball against a wall with my left foot. I never had trouble with the left foot again. Thanks to that frank comment I learned that confidence was within reach. I only had to recognize my failures and then apply myself.

Like an object with inertia in Newton's Law, my lack of confidence needed an external force, that chance remark, to act on it. I realized that I needed not blind approval but objective criticism, guidance and encouragement if I was to develop self-reliance.

During my two years at Downside I received much approval, encouragement and positive comment about my progress at the school. In all this I found motivation to do well. It seemed as if my teachers, in giving me praise, had become my coaches, my trainers, fellow participants in my effort. The selection to the school tennis team, the academic prize I received did much to remove my sense of insecurity.

Alas, all the kudos I collected went for nought when I finished school and moved to London. Whereas at school the end-point was the final set of examinations, clearly in view, in London I was alone again and the end-point was distant and not at all clear: what was to be my role for the rest of my life? It was a daunting prospect. Worse, I had no one in whom I could confide, no committed friend through whom I could see myself. I was spinning my wheels again, making no forward progress.

Stephen, my only point of reference in London, had no interest in what I was doing. His lack of approval was stifling. Always ready with scathing criticism, he never uttered a word of praise, even if I did something well. I felt like an actor playing to an empty theatre. It was like looking into a mirror and seeing no image. I felt adrift on an ocean with no hope of landfall. I needed blind faith in myself. Any meaningful achievement I might attain was still far away. Confidence was in short supply.

I took stock of my situation. Having been chosen by my classmates in Hungary for the scholarship to Downside was a sign of approval which I had earned. Although I arrived in England with no English I not only survived the tough immersion in a boys' school but performed actually better than some of my new classmates. This had to mean that I had what it took to pursue an academic goal, no matter how remote it seemed in the context of working as a copy boy at the advertising agency. When I asked myself whether my work ethic would see me through a difficult road ahead, I no longer had doubts about myself. It was an uphill climb but I had made my commitment. I still depended on Stephen but had to let his abuse go over my head and not lose sight of my objective.

As for encouragement to buttress my self-confidence, I just had to do without it.

Speaking in public did not come easily to me. During my boyhood in Hungary, I was swimming upstream and had not achieved anything of note to boost my morale. We were poor and in that sorry state confidence is in short supply.

My only experience of speaking before a group came from the method of teaching at my school where, instead of writing essays, as was the case in England, we were called upon at random to give oral answers before the class on the particular subject under study. At times we had to recite poems in front of the boys and everyone present was welcome to offer a critique. You never knew when you were to be called upon, so the secret was to be well prepared. The marks were "1", "2", "3" and "4", the last the failing grade. Our teachers kept a handy little book in which they recorded our marks. Some of the boys made a business out of noting down everyone's marks to the point where, towards the end of the year, they could present everyone, at a price, with the final marks they could expect. They were enterprising. I was not.

I spoke no English when I arrived in England and, although I was able to converse after some six weeks and received unexpectedly high marks on my written essays within three months, I would have been reluctant to speak in public due to shyness born of insecurity; insecurity, lest my diction evoked an impatient or impolite response from the listeners. By the time I had finished school, worked at a few jobs in advertising, store-keeping and as a wine delivery man, my confidence in verbal contact with others reached maturity. When I finished medical school I had reason to believe that I had passed hurdles where others would have fallen. I had proved myself to myself and if anyone should fault me for speaking English with an accent he would have to show credentials for sitting in judgment. After all, Einstein had an accent too. Interestingly, I found in life that those who would bring up the subject of my foreign accent spoke no language other than English.

Still, I had not been called upon to speak in public. While serving in the Canadian Army I was directed to give lectures to young soldiers on subjects of my choice. I suppose this was a way of filling in their time between drill on the parade ground and shooting at targets on the firing range. The lack of specific themes bothered me and I grew apprehensive until I remembered the lesson I learned in Hungary: the secret was good preparation! Thereafter it was easy. I chose topics I felt would catch the attention of 18-20 – year olds, made headings, wrote cues and practised delivering the talk in the time allowed. The boys actually listened and only a few were asleep while I talked. There was no standing ovation but I had broken the ice. Next time it would be easier. I did not realize then that there were many "next times" to come.

A short while later I had occasion to meet the Garrison Commander, a man whose strong glasses and the epaulettes declaring his rank signalled distant authority. Underneath it all he was the most kindly man, with a range of conversation transcending the accustomed perimeter of the conventional military mind.

"I sure enjoyed the lecture you gave to the boys!" he said in greeting.

"But you weren't even there, Sir".

"That's what you think! I was listening outside the door".

What the old colonel did not know was that his casual remark, his approval of my foray into hitherto new ground, was just the encouragement I needed. He gave me wings when I thought I had none. I forget his name, I do not even know if he is still alive, but many a time in later life when I looked at the expectant faces of a new audience I felt his presence, his kindly nod, his assurance that I had earned my way to the microphone and had paid my dues.

"Thank you, Colonel......"

For a long time I hesitated to write about this need for approval, thinking it was a defect in my character, that it would somehow diminish me in the eyes of anyone who became party to this revelation. Maybe I thought that being entirely self-contained, not needing the perspective of others, was a sign of maturity, a proof of "having arrived". Now I view it as a trait bordering on arrogance. Like it or not, we have an interdependence with our fellow humans, we owe it to them to fit in, to be part of the whole. We are not qualified to be our own sole arbiters without risking credibility. Truthful give-and-take will pinpoint our position and guide us to a better destination. Approval is a spur which justifies the agony of effort which earned it. We do need that image in the mirror.

Some years ago I joined an organization in England which offered critiques for aspiring writers on their efforts. At once I must make it clear that I do not consider myself a writer because I firmly maintain that designations of "writer", "poet" and "artist" should be conferred only after a significant body of work has earned acclaim. Still, I submitted some of my early writings and was assigned to my mentor, the playwright/actor Peter Whitbread. From his constructive comment and encouragement I drew new motivation for putting on paper experiences which I felt compelled to record for my children and grandchildren. Paper is a good listener and I had much to say.

As fate would have it, I suffered a heart attack shortly after my distant association with Peter Whitbread began. Stricken as never before, I informed him of my predicament. I lost the zest for writing for a few years. The need to unload later returned and I spent much time in revising everything I had previously put on paper. Seven years were to pass before I was ready for Peter again. My enquiry with the organization revealed only that he was no longer with them. I turned to the Internet where I learned that on a dark night some five years before he was knocked down by a car near his home in Norfolk, England and died.

My sense of loss was profound. With my focus now on writing I had lost the mentor who held up the mirror and let me see myself as I was.

I have reached a time in life when my own experience, my amassed knowledge, my humility will have to interpret me for myself, aware that inspiration to write comes with the need to judge, erase and try again. The temptation to look in the mirror is always there but the mirror may be gone.

CURTAIN CALL

I left the Iron Curtain behind me, with relief, in 1947. It was not without trepidation that I re-crossed it again some thirty years later. I had no choice, a family emergency compelled me to go. The thought that minefields, watchtowers, searchlights and trigger-happy border patrols stood in the way of anyone wishing to leave did nothing to increase my comfort level. The Canadian Pacific Airlines jet was full when I left Vancouver. A thick fog blanketed the continent of Europe and forced us into Düsseldorf, the only airport where we could land. The connecting flight to Munich was half full. On the flight to Budapest there were only six passengers. My pulse rate was high when we landed. The sign on the airport building left no doubt that I had arrived. If there was still any question in my mind the sight out of the window convinced me: a soldier toting a machine pistol at the foot of the aircraft stairs. Exactly what were they afraid of, I asked myself on the way to the terminal.

More guards with automatic weapons inside. I longed to see a London Bobby with good humour his only weapon. No such thing here. The passengers were corralled to a narrow wooden chute, one person wide, with an armed guard at each end, the booth of the immigration official in the middle. He sat behind a high barrier, only his eyes visible to the passenger. I handed him my passport. He looked at my face searchingly, then went to work on his gadgets, hidden by the barrier. A minute, maybe two, passed, while he looked up at me several times and examined my face for long seconds. Not being used to this kind of scrutiny I leaned forward and asked him: "Is there a problem?", looking him in the eye. He did not respond and went on with his work. At length he gave me my passport without a word. I was free to go.

On the way to my hotel I was struck by the multitude of communist party slogans on red background in the streets and the number of red flags. More shocking was the occasional sight of the national flag of red, white and green with a red star in the middle. The old-world hotel in which I stayed still looked the same, the demeanour of the desk staff polite and apolitical.

Still I felt unsafe. What if they did not recognize my Canadian citizenship and kept me in the country against my will? I telephoned the Canadian Consulate with my concerns. They were reassuring and told me not to worry. They said the transportation system was excellent, the food likewise and wished me a pleasant stay. Before they hung up they advised me to call any time I needed help. I memorized their address. I was on my own.

The thick fog which covered Europe descended as night fell, concealing all familiar landmarks. Not that I had trouble finding the family address, but I had hoped to look on familiar sights along the way. Although a fair distance away, this was a time to go on foot. The sidewalks and the cobblestones were the same as I remembered them. Only colour was absent. Everything was a dirty, dull, depressing grey.

The family reunion was an emotional affair. Those still living had all assembled, all loving, all pleased to see me. In my memory they had all been younger. Looking at them now, frail, with white hair, I became aware of the passage of time; time, which, no matter how joyful our reunion, will never be recaptured. Behind their smiles I saw the shadows of privations, fear and sadness which had been their burden through the years. The feast they prepared for me was beyond their means. Conversation was lively once the tears dried. Some of the talk still followed in whispers. The walls had ears. The informers who leaned on the wall beside the window during the war and the decades that followed still had their job to do. Only the faces differed. Nothing had changed in almost forty years.

Instead of returning directly to my hotel I first walked the route I took to school as a boy, in my hand a plastic bag with the works of one of Hungary's great poets in two volumes. The fog, very thick by then, hardly penetrated by the occasional street light. There was no one in the streets. I felt like a ghost. It was decades since I followed the route. Memories came rushing at me and I was heedless of the passage of time. I felt like the little boy of so many years ago passing houses I remembered, houses with their lights dead, their occupants unaware of my passage.

I passed by the arcade of the Corvin cinema, site of a deadly battle in the crushed revolution of 1956. The fog swirled among the columns as I passed. A human figure suddenly emerged and just as soon was gone. I saw the machine pistol across his chest as he passed. A voice rang out.

"Stop and come here!"

I stopped and turned but did not move closer. He was just near enough for me to see him.

"What are you doing here?" he snapped.

"Having a sentimental walk. I lived here once".

"Your passport, please".

"The hotel kept my passport".

"Your identification, then ".

"Where I live we don't carry identification".

"What's that bulge under your raincoat?"

"My camera". I unfastened the top buttons to show him.

"What is in the plastic bag?"

"The bag contains the works of the poet Mihály Vörösmarty, the complete works". I lifted the bag to give emphasis to my statement.

He turned on his heels and vanished in the fog.

I worked my way towards the hotel. It was still a long way and by now well past midnight. I must have been the only pedestrian abroad in the darkness, for I heard no footsteps, saw no shapes. Suddenly all the lights went out. A power outage with total darkness. There were no emergency lights, no candles in windows, there was no traffic. I was truly on my own. Having walked everywhere as a child I still remembered all the roads and side streets. I touched the buildings as I went and when I came to the end of each block I knew exactly where I was. I just had to feel my way along. At long last I came to the road which would take me to the bridge which, in turn, would lead me to my hotel. I strode out confidently in the pitch dark until I collided bodily with a figure going in the opposite direction. My hand felt the machine pistol slung across his chest. It was time to be friendly.

"A fine night to be out walking!" I called out.

"A fine night", he replied.

"How long before we see the lights again?", I asked.

"Only the good God's ********** knows the answer!"

His obscenity told me that he meant no harm. We called out a friendly "good-night" to each other, both already swallowed by the fog.

As I gained the bridge, sounds of drunken merrymaking reached my ear from the river sailors' pub below. A voice, well modulated by alcohol, speaking from the heart and dampened by the fog called Stalin a variety of names, none in praise.

A moment in history: the voice of the oppressed enjoying free speech.

GHOSTS ON A STAIRCASE

One interesting phenomenon of the immediate post-war period was the thriving of the businessmen. They all seemed to have food, comforts, money. Many members of society decided that the answer was for them to become businessmen themselves. Suddenly everyone had something to sell. Speculation, bartering, search for profit became the order of the day. Of course this was based on the false presumption that everyone could turn into a businessman. That this was not so soon became evident. Businessmen are born, it still appears so to me. To me the ideal businessman is the Jew. Sure, he is after profit, like everyone else, but he does not drive merciless bargains because he knows that a deal is a good deal if both sides are happy at its conclusion. A satisfied customer will return for more business. Those who try to make a killing on each deal drive potential customers away. For this reason I am happy to do business with a Jew. Business, and his ethics, are in his blood.

On a trip to Budapest in the mid-nineties I took my youngest daughter Zsuzsi to the old Jewish quarter of the city. We walked up the stairs to the top floor of Gozsdu-udvar in the old ghetto. I then told Zsuzsi to imagine what people there went through when in the middle of the night they were shaken out of their sleep by harsh commands and told to gather their things and be down in the street in ten minutes, or else they would be shot. Ten minutes, hardly enough to get dressed, let alone have the presence of mind to pack under threat. I told Zsuzsi to imagine herself walking down those stairs in the dark to an uncertain fate. I, too, felt the immediacy of the threat, the fear of uncertainty, knowing that the only certain outcome was loss of possessions, loss of family, maybe even loss of life itself.

Walking down the same stairs where they all walked to their fate, putting ourselves in their shoes, was a moving experience.

Just around the corner, in Dob-utca, we found a little shop at the bottom of some steps. As we walked in we found an old Jewish man with a friendly smile behind a makeshift desk. The sparsely stocked shelves were filled with Jewish religious artifacts. Half apologetically, knowing we were unlikely to buy anything, I turned to the old man and told him we would like to look around but I was not sure we would be buying anything.

"Look around!..........Why not?" he asked, raising both shoulders in acquiescence. "He who merely looks may become a customer!"

SOLDIERS OF THE QUEEN

In the Army we spent six weeks of every summer in field training at Camp Wainwright in Alberta. The camp occupies an expansive stretch of countryside traversed by the winding Battle River, rolling terrain, desert-like badlands with sparse vegetation, copses of poplars, idyllic ponds surrounded by tall grass and giant tiger lilies, an ideal setting for the hide and seek of make-believe warfare on the ground. In what once was a nature reserve, there is a resident population of wildlife, lynx, bobcats and owls, even an elusive cougar with which I once had an all too brief encounter. The sky, a big blue hat above this pastoral landscape, stretches to the far horizon in every direction. The light, the clouds and the moods of the weather are an ever-changing picture gallery to delight the eye, to inspire the soul. At night the vault above is lit by a myriad stars, so close you can almost pick them like fruit from a tree. In the moonlight field mice dart in and out of your tent, sometimes taking detours on your sleeping bag while you wait for sleep to find you. Now and then the heavens, aroused to anger, will shake you from your slumber with savage gusts of wind, you hang on to your tent till your knuckles bleed, while bursts of lightning create nightmarish silhouettes like the flashbulbs of an army of crazed photographers. You put your foot on the ground to steady yourself only to step on cactus needles which you then try to pull out by the obliging assistance of lightning flashes.

One sunny afternoon I was out in my jeep, exploring the fascinating countryside in which our training area was located. As usual, I was driving while my batman/driver sat in the passenger seat, searching for music on his radio. A station to our west reported a fast-moving storm heading in our direction. The sky to the west was indeed dark, bordering on black. We were far from our camp under canvas, so it was wise to turn for home. I was hoping to get there before the storm hit. I underestimated the fury of a prairie storm. We were in exposed terrain with sparse vegetation, cactus and stunted trees when the wind hit us head-on. I had never before experienced such fury. The velocity of the wind was such that I had to gear down, lest the jeep stalled. Riding in an open vehicle we felt the full force of the elements. I told my driver to huddle down on the floor and cover himself with a parka, while I continued the uneven contest with the storm. Heavy rain, driven by the fierce wind stung the exposed skin on my face and my arms. This was becoming an unpleasant afternoon drive. The sky above turned a black hue I had never seen before. Then, without warning, lightning was beginning to strike all around. A small tree, about a hundred yards ahead was hit, dinner plate-sized glowing cinders floating to the ground.

At this point the outing ceased to be fun. There was no shelter for miles around. Like persistent artillery fire, lightning struck now here, now there, close enough to make me wish I was elsewhere. I reasoned: we are in the only metal structure for many a mile, therefore I could expect to be hit. There was no point in stopping, so I just drove on in low gear. Soaked to the skin by now, the wind made me shiver.

Another thought: the jeep had rubber tires, maybe that would discourage lightning from hitting us. Soon I realized that the jeep, the tires, the ground were all wet, so there was no insulation there.

Worse was to come. The black cloud above discharged a barrage of golf ball-sized hail stones, clattering on the hood. This was serious, a man could get injured here! I reached for my steel helmet stashed behind my seat and wore it for protection. In all my time in the Army that was the only occasion when I had use for it but its presence on my head gave me comfort.

While all this was going on, my driver under the parka was playing pop-music from a nearby country radio station. He missed a marvellous spectacle. The storm was easing now, but perhaps its last hail stone hit the spoke of my steering wheel and ricocheted with painful impact on the cartilage of my ear. I suppose that is what they mean by a "parting shot". We got back to our camp none the worse for the experience.

The enemy for our war exercises was provided by visiting battalions from Britain, so similar to, yet so different from our own soldiers. Not only did they speak English with delightful articulation which Canadians called "an accent", but they took their task seriously. No more the "let's get this over with and have a beer" attitude of the easy-going Canadians, they came to play war and were earnest about it. When they took prisoners they made them take off their boots and socks and marched them off barefoot on sand, rocks, cactus and all. The temporary ill-feeling this created was ideal to create a combat atmosphere in which the prospect of capture no longer offered temporary escape from the drudgery of being a foot-soldier.

Armies have to eat and in the Canadian Army we were fed well. Our regimental corporal cook was a sensational chef whose influence on morale surpassed the impact of the most inspired motivational speech by any general. Away from our families and from the comforts of home, visits to the mess were the motivator, something to reach for each day.

The visiting battalion also enjoyed the boon of a Canadian Army diet, which only helped add vigour to their combative spirits in the field. Because they came to us without their field kitchen equipment, we obligingly invited their cooks to come over and learn to use the Canadian field stove, a piece of marvel in its day.

Bert, the English cook, was welcomed with customary Canadian warmth to our kitchen in the woods. The boys were busy peeling potatoes, a big job when you consider the effect of outdoor living on our appetites. Soon Bert was sitting in the middle of a bunch of jovial army cooks, all eager to know about his many travels in Her Majesty`s army uniform around the globe, especially as his appearance betrayed the fact that his boots had left their imprint on the soil of many a land.

To everyone's disappointment Bert was the laconic, silent type, totally lacking in the volubility the situation required. In a good-natured, convivial setting his was the only stern face. Still, the boys, many of whom had never travelled beyond Alberta or Saskatchewan, were eager to hear the travelogue of so travelled a man.

"So where have you been with the British Army, Bert?", asked one. "Fookin' Berlin".

"Berlin?!", retorted the questioner excitedly, hoping for a narrative of a devastated city, maybe Hitler's grave and juicy bits about the Cold War. "Tell us about Berlin!"

"A fookin' pile of ruins". As for Berlin, that summed it up.

The boys were not satisfied. This man had to be de-briefed before he would take all his exciting experiences with him to the grave. There were still lots of potatoes left to peel.

"Where else, Bert?"

"Egypt".

Now here was the prospect of real excitement, with the Sphinx, the pyramids and maybe veiled belly dancers doing their bit with the sands of the Sahara in the background. Bert remained silent. He had to be milked.

"Come on, tell us about Egypt!". It sounded more like a demand than a request. "A fookin' pile of sand".

The boys fell silent. There was no travelogue here, no narrative to lighten the load of what was still an alarming mountain of potatoes.

Time to change gear, thought another cook, not deterred.

"Tell us about the Queen, Bert!"

For the first time, Bert became animated. His eyes lit up. Flushed, he volunteered:

"The Queen?!....they should fookin' shoot her! Costs too much money, she does!"

"We're soldiers of the Queen, me lads..."
(Army marching song)

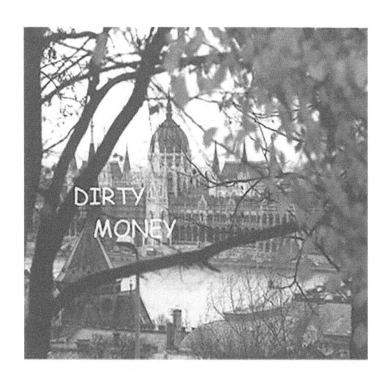

DIRTY MONEY

A black market currency deal is a cut and run affair. The trader must stay ahead of the police. He emerges from the crowd, strikes and vanishes without trace. His clients often end up his victims.

I always ignored the black market on my travels. What fiscal advantage might accrue from such deals would not even begin to compensate for the loathsome prospect of arrest and, God forbid! - incarceration. Languish in jail and return home in disgrace? Not me! I always shooed them off, "No deal!"

Always, that is, until I met the old man behind the Iron Curtain. And he was no black market trader.

Walking among the ruins of a historic site, the strains of a long-forgotten song caught up with me. I stopped and listened. Nostalgia swept over me and I set off in the direction of the music. There, among the ruins, in an empty shed, an old gramophone was playing a 78 r.p.m. record. I leaned on the door and waited for the song to end.

The record finished playing and was still turning around and around when an old man appeared. A little hunched over, gracious and friendly, he treated me as he would a daily visitor. We talked. He played the music for me once more. I listened in silence, thanked him and turned to leave when he, hesitatingly, called out:

"Would you sell me some American dollars?"

""I don't think so," I replied. "For one thing, it's illegal; for another, this is the last place I'd want to be in jail."

"Oh no," he said, "no one will know. I'll give you my money, you give me yours and we go our separate ways."

"Really, no thanks, it makes me feel uncomfortable."

He came closer and looked me straight in the eye.

"I need dollars," he whispered, "my only daughter lives abroad and I must see her before I get older. I could never get dollars through legal means. You are my only hope!"

I weakened. His eyes were kind, his voice sincere, his motive reasonable. I was grateful for the song he played for me. Should I refuse him? Was there really any danger? He saw me hesitate.

"Look," he pleaded, "the official exchange rate is forty. I shall pay you fifty.

Sell me thirty dollars, please!"

"I want no profit. I'll give it to you at forty, let's be quick about it!"

A glow of adventure pervaded me. Never before had I done this, but now I was not only doing it, but doing so in a good cause. We traded the money, shook hands and went our separate ways. The old man was playing the record again as I walked away.

By the time I reached downtown the temperature was in the thirties. From the canopied oasis of a sidewalk restaurant I watched the passing crowd while sipping my beer. The sunny side across the street was, deserted, everyone sought the shade. My insecurity of an hour ago had vanished. I had not been arrested nor followed. My freedom was not in jeopardy. I felt good.

"Want to sell me American dollars?" whispered the trader as he brushed by me on the sidewalk. He was clean-cut, young, with close-cropped hair. "Fifty-seven to the dollar."

This time I did not brush him off. Fifty-seven was a bargain. I did not fire a shot across his bow. He knew that I knew his rate was good.

The heat was oppressive, the sidewalk crowded. He stuck by my side. I had let my guard down earlier and now was doing it again.

"This is no place to do it," I said, implying that the deal was on. "Not with a crowd looking on."

"What about that doorway?" he asked, pointing.

"Won't do – too exposed!" I now realized that I was taking a considerable risk. If the black market in currency was illegal, the police must be looking out for it. The traders will be under surveillance and in their company so will I be. I brushed off the thought. My experience with the old man made me less anxious, even confident. The transaction had to be fast and out of sight.

"The side street", I said. "Just walk with me." In the side street I knew a U-shaped shopping arcade which may be less busy and which offered an escape if things got bad.

"It's a long way," he said impatiently, but kept close anyway.

We turned the corner. As expected, the side street was quiet, but being narrow, much hotter. We entered the arcade. He reached for his money.

"Put it away," I snapped. There were people about.

"It's OK," he replied soothingly, money in hand for all to see.

"We must find another place, this is not safe," I asserted and continued walking. We were back in the street again.

If there were any police about, surely we were spotted by now.

The obvious venue seemed to be an apartment house close by. We went through the front door into the lobby. At the far end there was an elevator with a stairway around it, three flights to each floor. I walked up one and a half flights, just to be out of sight of the front door behind the elevator. Out came his money again.

"How much?" he asked. "One hundred."

"That's five thousand seven hundred!".

He pulled out a wad of banknotes, folding them in a strange but deft way between his fingers and began the count.

"Five hundred, one thousand, one thousand five hundred..." He counted very slowly, finishing each number with a loud snap of the banknote between his fingers.

"Two thousand, two thousand five hundred..."

Just then footsteps approached rapidly through the hall and started up the stairs. A young man, about the age of my friend, walked by us and stopped within sight by an apartment door on the first floor. He appeared to have pressed the doorbell but the door did not open. He waited. Was he a policeman?

"Three thousand..."

"Shut up", I choked him off. "Put the money away."

"It's alright," he said calmly, emphasizing the last word, still holding the money in his hand.

I felt trapped. It was obvious that the young man on the first floor was not going to enter any apartment. He must be a policeman. Maybe this was a set-up. I was just about to call the whole thing off when he started down the stairs and walked by us quite slowly. Strange, I thought, he looks the same as my trading partner, the same age, dressed much like him. He disappeared through the front door. We were alone again.

My friend saw my discomfiture and struck a cheerful note.

"No need to worry, there are all sorts of people about, but they are no danger to us. Let's start again!"

"Quickly," I said.

"Five hundred, one thousand, one thousand five hundred," the count was painfully slow again, "Two thousand, two thousand five hundred, three thousand..."

The front door opened again. Brisk footsteps approached and started up the stairs. The money was in the open. There was no time to conceal anything. Another young man, much the same age, dressed the same way, walked by us and stopped on the first floor.

"It's over - let's go," I said, starting down the stairs. The pressure was building. The slow count, the two young men making pointless visits to the first floor, the risk, the heat, the mounting fear about a trap: it was too much.

These unwelcome visitors must be the police, I reasoned. Two of them had witnessed our clumsy, clandestine deal. Of course, I had not parted with my money yet, so maybe I was safe. Then again, why was the dealer so complacent? Maybe he was in league with the police! I was losing control of the situation. This was more than I had bargained for. My instincts said, "no more!"

The dealer became overly friendly and reassuring. He put a hand on my shoulder.

"Don't worry about them! They are nothing. Let's go next door and close the deal quickly. Okay?"

Backing out is not something I do easily. Even in this illegal set-up it seemed the dishonourable way out. Maybe I was overreacting, I allowed. I had agreed to the deal, now I must carry it through.

The apartment house next door had not yet recovered from the war. The front door opened into a narrow lobby, more of a corridor, with the apartment of the concierge at the far end. To the left a door with frosted glass led to the staircase and the elevator. We stepped in and closed the door.

"This time be quick!" I snapped, reaching for my one hundred dollars. I sensed something was very wrong. I was at a dead end with no chance of escape.

The ritual began again. The count a little quicker this time but still painfully slow.

"Four thousand, four thousand five hundred, five…"

I heard the front door open and footsteps approach down the corridor. A figure stopped by the frosted glass window and a face was pressed against it – a grotesque glimpse out of a nightmare – it stayed pressed against the glass for a few seconds, then withdrew. The footsteps receded and the front door closed.

"Five thousand five hundred. Sorry I don't have bills of one hundred. I have to give you two hundred in tens."

With this he pulled out a bulky wad of greasy, revolting banknotes and counted off the two hundred. He held out his arm wide to one side.

"Five thousand seven hundred!", he said. "Now you give me the one hundred dollars."

We exchanged moneys simultaneously. The touch of the greasy banknotes filled me with revulsion. There was no relief now that the deal was over. My friend stuffed the dollar bills in his socks and advised me to do the same. I just stuffed the notes in my pocket and opened the frosty glass door, wondering what was waiting for me outside.

"I'll stay behind and let you go first," he said.

I opened the front door and stepped out into the heat. Leaning on a shop window opposite was one of the young men who paid a visit to the first floor one building ago. He looked at me with folded arms.

At this point I knew that I had been set up. I must get away, I thought and walked briskly into the busy shopping street round the corner. I crossed and recrossed it, backtracked, watched the reflections in shop windows for signs of being followed. The heat was even more punishing now. I felt like a hunted animal. My heart was racing. In my pocket: the evidence.

At last I found myself well down a quiet side street, Aranykéz-utca, the "Street of the Golden Hand" where goldsmiths had once plied their trade. I turned. No-one followed. I heaved a sigh and collapsed into my rented car in the parkade.

I was safe for the moment. Surely, this must have been a planned operation. The slow count, the interruptions, they must have been designed to put pressure on me. Still, I came through and that was good.

But then who were all those people? Does it take three men to secure one deal? That would not be profitable. Perhaps they were police after all………The dealer's brazen approach in a busy street frequented by tourists must be obvious to them. In a country short on hard currency they would keep an eye on such things! If they were police then I was far from safe!

I shifted into gear and drove off, still thinking. If they were policemen, why was I not arrested? Perhaps I was being followed. I looked in the mirror. There were several cars behind me. They all looked the same, as is usual behind the Iron Curtain. One was a cream-coloured *Lada*, the others grey and blue. I drove around, making frequent turns, now right, now left. The cream-coloured car stayed with me for a while but when I looked in the mirror again, it was gone. Finally, no-one was following. I was safe, after all.

The uncertainty, the tension and the heat began to take their toll. I stopped for lunch in a quiet restaurant. The pace was suddenly slow and comforting. The waiter was polite and laconic, the beer cold and soothing, the food incidental. I went there for refuge, that's all.

As the minutes passed and nothing untoward happened my composure slowly returned. You stupid fool, I chided myself, getting into a scrape like that for a cheap thrill! I shrugged and pulled my ill-gotten gains from my pocket! Surely, I could stop worrying now!

The greasy, filthy banknotes were repulsive to touch as I began to count. But what's this?! The five hundreds were nowhere to be seen. My heart sank. I tried again: just dirty bills of tens, nothing else. That's it! The strange grip on the banknotes and the deft flip of the fingers had done their job. I had been duped!

An unexpected calm descended over me. The sleight of hand which parted me from my hundred dollars also reassured me that the game was over, and with it, any fear of police action. The dirty banknotes, designed to put me off were, in fact, my salvation. I had paid for my lesson in full and was free to go. I tossed the money in a waste basket on my way out.

A few days later, at the airport checkpoint, the official looked me in the eye and asked "Did you change money on the black market?"

"Only fools do that!" I spoke the truth.

He let me pass.

JANUARY RIVER

I always harboured a secret yearning for a far-off destination, alive only in my imagination, where troubles vanish and dreams come true. "To see Naples and die...", went the old saying. Forget Naples! It is a tired prescription. Besides, if you are a pessimist you could literally interpret a fiery end for yourself in case Vesuvius chances to be in a cussed mood during your visit. Not for me! It is Rio, Rio de Janeiro that held my imagination captive.

When Portuguese sailors entered Guanabara Bay on January 1, 1502 they believed they had found the mouth of a large river. Mindful of the date and inspired by the imbibitions of the night before they christened it Rio de Janeiro. January River. Though there is no river, the name stuck. Some sixty years later a city was founded which still bears the name.

Just to see *Rio de Janeiro* written on the board at the PanAm departure lounge sets my heart beating faster. At the end of the long night flight I shall walk the streets, inhale the air, feel the heartbeat of the city of my dreams. My feet will tap out the rhythm of some Latin American song and I shall smile back at the smiling faces that await me. A prospect filled with expectation.

A glance at the passengers around the room has me puzzled. Smiling faces? There are none. They look disinterested, tired, sullen. Where is the zest that I am feeling? Where is promise of laughter, gaiety and music? Their faces are blank, they seem to stare without seeing. Figures in a wax museum.

I, on the other hand, am busy surveying the departure lounge and its growing population with concealed interest. I recall the saga of the Uruguayan rugby team whose airplane crashed on a high plateau in the Andes mountains of Chile. The few survivors had lost all hope when the search was abandoned. After weeks without food they themselves faced death by starvation and exposure until one of them proposed that they should eat the flesh of their dead friends. "Did Jesus not say: this is my body, eat it and you will have………life!" – was his argument. Reluctantly, they gave in to cannibalism and lived. It is in me to look for falsehood in righteous indignation and there certainly was plenty when the news got out. Those who turn in moral revulsion from the prospect of eating human flesh should experience the stark reality of being cut adrift with no prospect of rescue, watching friends die one by one, under attack by the elements, their bodies digesting themselves from within just to hang on to life a little longer, their last meal only a fading vision while vultures circle above, waiting for their next.

As a survival drill, I pay close attention to the bodies drifting into the departure lounge. Who knows? You know…… must I say it? That big fellow coming in just now, he has lots of muscle but ah, look at the fat around the waist! Bad for my arteries!……The slim middle-aged woman was a good prospect until I saw smoker's wrinkles on her face……Now that pretty girl over there is something else but please God, don't let her die...!......And so I spend the time, no one aware of what I'm contemplating.

To enter the departure lounge we must pass through the metal detector, a reminder that some passengers may have in mind a destination other than the one printed on my ticket. The world of the nineteen-seventies is upon us!

Just now a swarthy fellow comes in under the metal detector, setting off a concert of whistles and beeps. The beads of sweat on his face would more than suffice for a rosary. His big stomach hangs apron-like below his belt. His teeth are uncared for. Definitely not foodstuff. His T-shirt, all too tight, is drenched with sweat, clinging to the flab overflowing his trousers. The trousers are so tight I wonder how he ever gets out of them. His side and hip pockets bulge with unseen contents.

"Will you step this way, Sir?", he is made to spread his arms and legs while the detector wand sweeps over him. As each pocket is passed a loud metallic alarm penetrates the silence. By now no one talks in the room and all eyes are on the new arrival. He is led to a small table.

"Please empty your pockets, Sir", directs the security woman, pointing to the table.

A look of torment visits the man's face and he shakes his head in denial. The attendant remains patient but clearly intent upon seeing her instruction obeyed.

"Your pockets, Sir!", this time with emphasis on the last word. You could hear a pin drop in the room. The man, sweating profusely, is still shaking his head, his eyes begging to be let off. Passengers in the lounge lean forward in their seats, watching the drama unfold. The man looks around the room as if trying to recruit support, support to allow his pocket contents to stay where they are. No one rises to help. He is alone. Desperately alone.

"Your refusal to cooperate may have serious consequences! For the last time, Sir, please empty your pockets!" The tone is unequivocally serious. The security woman means business. She is good at her job. Still polite, but there is no mistaking her 'comply or else!' message.

The man, crestfallen, trapped, spreads his arms as if crucified, in abject defeat, the last moments of a cornered fugitive turning to face his pursuer. He reaches into his pocket and begins to place its contents on the table in full view of everyone present. With all his pockets finally emptied, their contents form a heap of considerable proportions. I doubt that anyone in the room, male or female, had ever seen such a pile of condoms. The wand is swept over them and the foil-wrapped contraceptives respond like the brass section to a conductor's baton. The security woman turns away in disdain. Crushed, broken, humiliated, the man fills his pockets again and retreats to a corner, slumps into a chair.

It is a good omen when the seats next to you are unoccupied as the airplane door closes. The Boeing 707 rises purposefully into the air. Filled with anticipation of what is to come I do not give New York more than a perfunctory glance. The airplane is only half full. The passengers are quiet and I am left alone with my thoughts. With only three days at my disposal in Rio I must use my time well. Who knows, I may never return.

No one knows that I am carrying a secret. I had given up smoking cigarettes some months before the trip but the undisciplined side of me insisted that if ever there could be dispensation from a vow so made, it had to be in Rio. Nursing my drink, well above the clouds now, I pull out my packet of Chesterfields and light up. A flush of excitement, the thrill of trespass out of bounds, visits me as I take puff after puff. Smoking a cigarette takes about ten to twelve minutes, a short time indeed. Yet, somehow the time it takes to indulge in my indiscretion now seems much longer. But why?

After the first few, carefree puffs my conscience touches me on the shoulder. I look at my smouldering cigarette as if it was a childhood friend with whom I was told not to play any more. The smoke fills my nostrils with a long-cherished aroma of tobacco. The grip on my shoulder tightens and I start thinking of all the reasons why I gave up cigarettes. I also recall the doubt thrown in the way of that decision at the time, coupled with anxiety about how I would manage without them. How will it change my daily routine, what about my morning coffee without my precious smoke? Or is it really the smoke that makes me drink the coffee? Then again, a cigarette is such comfort in moments of stress, moments when a decision hangs in the balance...or is it just a means of putting off that decision, any decision, a little longer?

Is it a friendly glow of encouragement from my cigarette that I now contemplate or a red warning light trying to bring me to my senses? I flick the ashes and take the last puff. I am not enjoying it at all. My little indiscretion was nothing more than a childish way of seeking excitement, knowing there will be no consequences. Somehow I am past that now, I am my own man, I make my own decisions and take satisfaction from that. I had agonized over that decision for months. Why discard it now on a passing whim?

Dinner arrives as daylight fades. With the airplane less than full the service is more personal, more attentive. My wine glass is refilled several times. I inhale the aroma of the coffee and soon temptation beckons again. What the hell?! I pull out my Chesterfields and light up. At once I realize that I am not enjoying it. Maybe I'm wrong. Another sip of coffee, another puff. Don't they go hand in hand? No, it's not working this time. Better admit to myself that I had made a mistake………I put out the Chesterfield and leave the almost full packet to be taken away with my dinner tray. The brandy and the wine make me feel sleepy, mellow. The last thing I remember is the enormous delta of the Amazon River lit up by the moon. I drop off to sleep and dream of what morning will bring.

My taxi driver in Rio believes that he is a reincarnation of the renowned Grand Prix driver Juan Fangio. So it appears as we head for my hotel along a palm tree-lined boulevard, three lanes in each direction. My heart in my mouth, I forget to look at the scenery, staring straight ahead, waiting for the inevitable crash. Tearing down the centre lane at about 50 miles an hour he threads the needle between a bus on the one side, and a large truck on the other, both stopped. As we hurtle through the gap a child appears running across the road ahead.

"Jesus...!!", I exclaim in anger.

The driver, still speeding, takes his hand off the wheel and points toward the distance. "Yes, Jesus, on Mount Corcovado!"

Indeed, in the far distance stands the statue of Christ the Redeemer atop his mountain, hands outstretched, in resignation it seems, because bringing order to this crazy traffic is clearly beyond his control.

My hotel room gives on to a fine view of the Sugar Loaf and of Flamengo Beach below. Before I even unpack I turn on the wall radio, eager for the sounds of South America. The first thing to reach my ears is the Portuguese version of the American advertising jingle:

"Double your pleasure, double your fun
With double-good.......Doublemint gum......"

In horror, I switch to other channels. More of the same. My first disappointment.

Never mind, soon I shall be sitting at some terrace, taking in the scenery, sipping Brazilian coffee! I set out early, determined to walk, as I do in every city I visit for the first time. The state of the sidewalk and the gutters surprises me but I am here to observe, not to criticize. Disappointment greets me again. Coffee is served only in little stand-up corner bistros, none of them too inviting. Ah, well! I remember not to drink the water, so I stop now and then to hydrate myself with bottles of "Brahma Chopp" beer.

The avenues are beautiful, tree-lined, the shade a welcome relief. People everywhere go seriously about their business. I notice a fascinating mix of skin colour and facial features, maybe a preview of our world in a hundred years. The gene pool has been generous to this nation. The lines on the faces of passers-by are not the footprints of smiles. Beggars with small children reach out a tired hand from doorways where they shelter until they are chased away. I hear no laughter, no song, witness no smile.

I go to the beach to escape the traffic. The stretch of white sand is wide, generous. For the first time I can admire at length the magnificent, almost arrogant beauty of the Pão de Açúcar, or Sugar Loaf, as we call it. Little cable cars carry visitors to the top. Soon I shall be one of them. Along the sand there are countless soccer fields where the boys play the game from early morning till eleven at night when the lights go out. Their skills with the ball are remarkable; no wonder this country does so well in international competition.

Vendors patrol back and forth by day, selling junk food to the sunbathers. At night large rats patrol back and forth, feasting on the leftovers.

Here and there I come upon beautiful homes behind forbidding fences, fleets of Mercedes cars parked within. The gardens are manicured, the trees healthy and proud, the flowers alive with vibrant colours. You get a good feeling until you turn your head and notice the *favelas*, wretched slums clinging to the steep hillside next door.

Entrance to the mansion is barred but I am free to enter the *favelas*. Being curious, I begin my ascent along the winding path among the tarpaper and corrugated iron shacks. Small children eye me hesitantly from behind makeshift fences. Skinny, sick-looking dogs look up but show no interest in my approach. Pigs, running loose, rummage in the smouldering heap of garbage by my feet. I feel alone until I come upon a smiling woman who looks at me puzzled, wondering what I am doing there. I choose to stop and attempt a conversation which goes smoothly despite neither of us being able to understand the other. She wants to know what I carry in my shoulder bag. I make a joke of it when I realize that I carry all my camera equipment and all my money in that bag. Still, she gives me the first Rio smile and that is worth something. She is destitute but she has an inner glow. I carry on our smiling chat a little longer, then give her a handful of *cruzeiros*, leaving her a little less destitute, only for the moment.

I keep climbing higher. The clothes people wear are little more than rags. Thank God for the kindly climate, for they do not need much protection from the elements. Fashion, good shoes, decorations for their pathetic homes, even palatable food and safe water are not on their list of life's offerings.

Along my incursion into the *favela* I become conscious of a recurring sound, a shout that follows me in my upward path. The source of the voice is never seen but it is obvious that my presence is being observed and the shout serves as a warning of trespass by a stranger. There is no roadway as such, there are no recognizable alleys even, just shanties nestled shoulder to shoulder. Beyond them: jungle. It occurs to me that if someone of evil intent chose to do away with me and take my bag I could be buried, vanish without trace, never to be heard from again. A sobering thought.

I turn to retrace my tracks. On the way down I meet more people, curious onlookers. No hostile faces, these are simple men and women whose only gift is their friendly smile. I greet them all, stopping here and there for a little "conversation". Strangely, I feel no threat, despite the obvious insecurity of my situation. In truth, kind words are all I can give them and I gain the impression that it is not an experience to which they are accustomed. They wave as I leave.

Have I just visited purgatory or was it hell, I ask myself, walking away. I have just seen a glimpse of humanity for which society appears to have no use, no concern; dregs of life serving no apparent purpose in the scheme of things. They exist outside the framework of community, unwanted, unprotected, shunned, with no one caring if they live or die. I wonder if they visit the dreams of the mansion-dweller next door.

I choose the wrong day to visit the Sugar Loaf. The line-up for the gondola takes two hours. I stay and observe the crowd around me. Young girls behind me chatter incessantly without meaningful content. Ahead, a dignified elderly woman is accompanied by a pretty girl of about sixteen. They talk quietly and keep to themselves. Then there is the gregarious one who tries to engage everyone in conversation until he finally gives up. The heat is punishing.

The trip is worth the long wait. Views of Rio are superb, wherever you care to look. I do not enjoy the trip down the flimsy cable and am relieved when I transfer to hard ground. I choose to walk back to Flamengo Beach around Botafogo Bay. The sun is getting low and a cool breeze is blowing from the sea. Traffic along the boulevard is quiet and the noise of the day is dying. Palm trees and flowers along the route remind me I have reached my coveted destination. I feel good.

Suddenly I notice a figure running across six lanes of traffic in the boulevard. I recognize her as the teenage girl from the cable car line-up. She is heading straight for me and before I can react she holds me by the shoulder, looks at me for a moment and kisses me on the lips. "I love you!" she cries and holds on for a fleeting moment before she runs away. I reach out, but she is already past my outstretched hand.

Has her kiss turned me into a *carioca*, as residents of Rio call themselves? My eyes try to capture this fleeting apparition, something to etch into memory, but all I can recall is her innocence and the touch of her lips that was mine for a brief, undemanding moment, a moment so short-lived that I am left without a face to recall. Her figure recedes. In no time she is back with her elderly woman companion, then I see them no more. Why she kissed me I shall never know, nor will she know that she gave me a gift for life. There is no name to remember, no identity, no permanence in the encounter. We shall never face nor hurt one another. The moment is eternal because it had no beginning and no end.

A tram car takes me to Mount Corcovado. The rails wind back and forth up the steep hill. A voodoo sacrifice, *macumba*, has been laid by the route in full view, a rooster with its throat cut, some symbols and flowers. The shadow of Africa persists in occult religious practices. It makes me feel bad. No living creature should die for nothing. The Christ statue above is so enormous that people beneath it appear like Lilliputians. The view is amazing but the one sight that stays with me is the appearance, from a great height, of Maracana Stadium, perhaps the largest soccer arena in the world. The way its oval roof surrounds the playing field below is reminiscent of some refluxing ureter openings I see in the course of my work. For ever after I would describe them as "stadium" orifices in my operation reports. Thank you, Rio.

A glass display case in the hotel lobby serves to advertise *The House of Stein*, a jewellery business with a fine selection. A limousine will take you there and bring you back. I make an appointment.

The driver drops me off by an imposing building with a sculptured front door. The bell is answered by a solemn, elderly man who beckons me to enter. An elegant middle-aged woman appears and gives me the most courteous greeting. Of all the people I have met in Rio she is the most refined. She engages me in brief conversation before turning to business.

"What is your wish, Sir?"

"I wish to see gold rings".

"Very well", she snaps her fingers and from the wings there appears a stunning, olive-skinned, dark-haired girl. "Show the gentleman what he wants".

The girl leads me into a salon where time stands still. A subtle light of afternoon filters through the curtain, pervades the room. She offers me a seat. "Would you care for tea?", she asks. When I accept she snaps her fingers and from behind a curtain there appears another, equally beautiful but much younger girl. After a brief exchange in Portuguese the young girl withdraws. The olive-skinned beauty speaks

English well and we have a delightful conversation on wide-ranging subjects. The tea arrives and we continue talking. At long last, she snaps her fingers again and the young girl returns with trays of gold rings. By this time my ego is so flattered that I know I shall pay whatever she will ask. I make my choice and we settle the matter. Her handshake is warm, lingering. She says good-bye. I leave the house flushed with good feeling. I have been seduced, I know, but nothing I had ever bought gave me such pleasure.

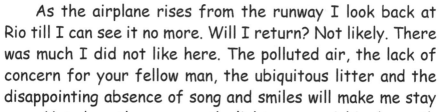

As the airplane rises from the runway I look back at Rio till I can see it no more. Will I return? Not likely. There was much I did not like here. The polluted air, the lack of concern for your fellow man, the ubiquitous litter and the disappointing absence of song and smiles will make me stay away. Yet, down there among the lights now vanishing beneath the clouds there is a beating heart, there is excitement, a different style, a way of life that stays with me. Raw truth stares you in the face: the good and the bad. If I did not like what I saw, maybe it is because I was not open-minded enough. The overwhelming beauty of this "River of January" comes out in triumph over the concerted attempts of Man to destroy it. Rio will be there long after we are gone. The Redeemer will watch it all without interfering with the pulse of this remarkable place. Girls will still bestow stolen kisses on complete strangers who, for those fleeting moments, will feel like *cariocas*, and then leave with memories that no man can take away.

SANDCASTLES

The best journeys are those undertaken on a whim. My trip was precisely that. Decide, pack and leave before doubts assail you!

I had been reminiscing about London and in a moment of clear vision I decided to hop on an airplane and meet it face to face after a passage of some forty years. No planning ahead, no phone calls, no tiresome travel arrangements, no time to change my mind. London had to take me as I was.

"The reason for your visit?" asked the immigration officer at Heathrow.

"Nostalgia", I replied.

"As good as any", she quipped, stamping my passport.

The gate to London, the door to my past, flung open.

The Piccadilly Line ride into town was a homecoming of sorts. The smell of the carriages had not changed, the faceless suburbs rolled by, the station names, like beads on a necklace, proclaimed their permanence. After some forty minutes the metallic rhythm of the rails had me feeling like a Londoner again.

Immersed in a bathtub so deep that only my nose was above water I began planning for the next day. Planning it remained because I only just made it to bed before I fell into a deep sleep. I did not wake until late the next morning.

The District Line train clattered past the old familiar stations…………………Temple………Blackfriars………Aldgate East I finally got out at Whitechapel. Thrilled to walk in my footsteps of old, I emerged into the sun of the Whitechapel Road. The noisy old street market was busy as ever, the traffic still a restless river, the passers-by still preoccupied. Across the road the dark shadow of the London Hospital, my *alma mater*, declared its authority. By now it had graduated to the "Royal London" designation.

I entered the hospital by the front entrance where in my houseman days ambulances would pull up with lifeless bodies. One of us housemen was summoned to declare the body ready for the morgue. It always gave me the willies because the clamour of the road and the rumble of the underground train below rendered checking of the pulse or the use of the stethoscope useless. Everything around danced to the pulse of Whitechapel. The fragmentation of blood vessels seen through the ophthalmoscope was the only useful sign but the man had to be dead for a while for that to occur. I always feared that someone I pronounced dead in the ambulance would later sit up in the morgue and ask for a cup of tea.

The porters inside were courteous but the atmosphere had changed. No longer the veterans of old, whom we treated as family, they were now the corporate types, impersonal, efficient, businesslike. Gone were the days when they doubled as security men, when the man on call might fail to answer his page because he was at home in bed with heart failure.

Hardly surprising that when I enquired about renowned physicians and surgeons of yesteryear they answered with a shrug. "Not on my shift Guv………", "now if you'll excuse me………". The guard had changed, the past had no relevance.

The Medical School library, where silence once ruled, science digested and priceless information gleaned, was empty, perhaps ready for demolition. Within its walls were held the Christmas shows where talented students took the mickey out of revered teachers who were flattered by the recognition so bestowed. A rather withdrawn, uncommunicative eye surgeon, a Mr. Lister, became a convivial, amiable man after being immortalized by the lyrics:

> "I had a job with Mister Lister, oh, yes!
> Mister Lister kissed a sister
> Sister ducked and Lister missed her.
> That's the job for me…"

I hummed the catchy melody of the song about Bertolini's and Murphy's, cheap cafés in the shadow of the hospital with their menus tailored to the meagre means of the medical student. It defined Whitechapel with its refrain:

> "Bertolini's at seven, a table just for two,
> My idea of heaven, spam and chips with you……

> *Murphy's on a June night, up above the stars,*
> *Turner Street in moonlight, full of stolen cars,*
> *Mem'ries of Whitechapel in the spring.........."*

The Medical School office staff conjured up three curled, faded photographs of Deans of the past. Beyond that they had no information on the teachers of old who in their day represented all that was the best in medicine.

The cold wind of four decades had swept traces of everything I held dear. The past appeared to have sunk into the quicksands of oblivion. At the fast pace of forward progress there was no time for a glance over the shoulder. I took a look around, knowing I would not return.

When I was a boy fresh out of school I worked for Smith's Advertising Agency at 100, Fleet Street. They styled themselves "The Firm in a Hundred". Thinking back on the lively pace of the advertising agency, the endless deadlines, the location right in the heart of the British Press, I hastened my steps as I approached the building. I even looked forward to a cup of tea, which I was sure to be offered, only this time it would be made by some new office boy, fifty years my junior. But what is this?! The once shiny windows looking down on Fleet Street were dirty, the building in a state of disrepair, the front door padlocked. High above, an unsightly sign declared that the building was for sale.

As for the agency, it no longer existed, its people were gone, the "family", as they often characterized themselves, scattered, maybe dead and buried. I stood, looking at the windows for a while, remembering those who had worked within, remembering myself as a young boy carrying trays of tea which it was my job to make. I recalled the kindly old spinster in the accounts department who would wag a finger and say "Nicky, tut-tut, you took the kettle to the teapot again, not the other way around!" I was sure that the old, chipped teacups were still gathering dust up there in the attic, the site of my apprenticeship in the art of tea-making.

I must confess, my first two visits left me depressed. The acquaintances I had hoped to renew did not materialize. Still, I decided I would return to the gourmet restaurant in Duke Street, attached to the wine business where I worked as a delivery man back in 1952. I could now afford to order whatever I wished, more than that, wash it down with fine wine and reminisce about the old days. Days more tough than good.

Duke Street had not changed. From a distance everything seemed in place, as if the street had been a monument. I walked with anticipation, already savouring the wine I would soon be sipping. I thought of Michael Broadbent, the owner's understudy, a polished, elegant young man who in time became a world authority on wine. I thought of Molly, the Irish cook who always had something savoury to say when I appeared. Already hungry, I resolved to enjoy the next hour.

The restaurant no longer existed. The wine business was gone. There was no trace of Michael Broadbent, of Molly or any of the staff who were part of my world all those years ago. Still hungry, I sat down on a bench in Manchester

Square to contemplate the passing of time. The past, was it more than a figment of my imagination?

I was on a mission and though I had drawn three blanks I was not deterred. London was an ocean, too big, too impersonal, people could vanish, leaving no trace. Had it not swallowed me like a whale back in 1949? Clearly, I had a better chance of retracing my steps at the hospital in Barnet, Hertfordshire, where I spent time as a surgical houseman. I would dearly love to connect with the surgeons with whom I had worked. Who knows, they may even be interested in meeting me again.

A gentle rain fell as I walked into the hospital grounds. Things looked much the same as before and I had no trouble finding an office where I stated the nature of my enquiry. The woman listened politely, with the trace of a smile as I asked about the two surgeons. She spent a moment trying to remember, then shrugged her shoulder in apology.

"Sorry, I have not heard of these surgeons! Just how long ago did you say you worked here?"

"Forty years", I replied cheerfully, as if it was last Christmas.

A smile came over her face. She shook her head and turned to a man nearby, repeating my enquiry to him. They both smiled, shaking heads. By this time I must have looked forlorn, for the woman suddenly gave me encouragement.

"The secretary in the postgrad office will help you, I'm sure. She has an excellent memory!"

Dejected at my fourth failure of the day, I stepped out in the direction advised. Gentle drops of rain on my forehead lent a soothing, consoling touch. Full of hope, I entered the postgraduate building and soon found the person I was seeking.

"I don't usually forget names, but I'm afraid I cannot help you"

Seeing my disappointment, she added: "Nobby is your man, Everyone passes through his department,……in time. Do go and see him!"

"And where do I find Nobby?"

"In the morgue, round the corner".

I thanked her, adding: "when you have dinner with your husband tonight be sure to tell him you have seen a ghost!"

The morgue was locked but the bell was answered promptly by a man, who, judging by the bloodstained rubber apron he was wearing had to be Nobby. He was bubbling with enthusiasm and before I could even state the purpose of my visit insisted that I follow him to look at the two new autopsy slabs used for the first time that very morning. His joy over the body-fluid- stained slabs was the only bright part of my otherwise frustrating day. He flushed the surfaces clean and with the satisfied look of a master craftsman hung up his rubber apron. He washed his hands thoughtfully and headed off in the direction of the kettle.

"You would no doubt enjoy a cup of tea". "No doubt!", I replied.

While the water was heating up I told him of my time at Barnet and of the people with whom I had worked. He listened with his chin on his chest, clearly placing himself in the coordinate system of time I was presenting. Even after I had finished talking he remained silent. He got up, made the tea and sat down again, both of us nursing the cups with their comforting, warm contents. Deep in thought, he was reviewing the parade of faces that had passed by him, doctors eager to find answers to riddles of pathology or expressionless cadavers lying on the autopsy slabs. My confidence grew. I knew that Nobby, this contemplative, "slab-happy" gatekeeper would at last lead me to the Holy Grail.

Finally he broke the silence.

"I have never even heard of these men".

Did I say "London had to take me as I was" when I set out? An arrogant attitude, in retrospect. As it turned out, London taught me to take things as they were. I was not the dominant figure in the equation. A trip undertaken on a whim is not entitled to expectations.

I was like a child who returns to the beach looking for the sandcastle he had built on holiday the summer before. I felt foolish, disappointed, humbled. My quest for the past was laughable. It did not deserve to succeed. At long last I came face to face with my child-like denial of the passage of time, my assumption that it was like a story in a book you can take from a shelf and re-live any time at will. Even if I had succeeded in bringing those old friends to life the circumstances and they themselves would have been different from the way my memory, my wishful thinking had preserved them. Said the Romans long ago: "times change and we ourselves change with the times". Why jeopardize in a moment memory of a relationship buried long ago?

The past is a ship that has sailed, a cloud that has gone its way, a bird that has flown. You cannot live again any part of it.

Who was it who said "*partir, c'est mourir un peu............*"? To go away. to part, is to die a little. With the act of parting the play, the interplay between us comes to an end. We say "au revoir" almost in denial, because it is easier than to say goodbye. The script has run its course, the curtain ready to fall.

Time is a continuum of moments, vying for our attention. We must pay heed in the moment, for time bestows no raincheck, no second chance, no privilege of a "time-out". Now it's here, now it's gone.

The future I once reached for
Has become the past.
The moment I now contemplate
Has already passed.

LOVE BETWEEN TWO ISLANDS

The best of dreams catch you unprepared, lost between fantasy and fact, when the things you only wish for suddenly appear tangible, the impossible takes form, hanging on the delicate thread of a passing moment. You reach out, want it to be not a dream at all, this splendid, evanescent eternity in the wink of an eye, this vision which yet eludes your touch, leaving you with a beating heart and an empty hand.

Dream or reality, it matters not. You are left with traces of a smile, echoes of a voice, which later may return in those timeless moments when the distinction between dream and reality hardly matters.

"The young lady will help you."

I walked over and faced her. She stood relaxed, graceful, confident and waited for me to speak. Her knowing smile assured me that I was in good hands. Like most Chinese girls, she had delicate build, black hair, immaculate skin and tidy eyebrows over widely set eyes.

The moment our eyes met I felt as if an elemental force had blown away the windows of my house and invaded its every room. I tried to speak but could utter only a helpless, stricken sound before I was able collect myself and state my case. She remained calm, attentive and showed no response. She helped me settle the business at hand.

I stared at her apologetically, still trying to find the right word. I wanted to say something to save myself, something plausible, something appropriate. At length I realized that nothing but a full confession would do:

"You are the most beautiful girl I have ever seen!"

The words just tumbled out. Was I addressing her or talking to myself? Did I even know what I had said?

She smiled. She thanked me. Our encounter was over.

I knew I would not rest till I saw her again.

The next morning there was a short line-up of people waiting to see her. I had no business to transact but stood in line.

"Yes, sir," she smiled expectantly when my turn came. Time to confess again.

"I just stood in line to say......good morning!"

The warm Singapore air outside felt like an embrace. Walking aimlessly, I did not know where I was. My day had already begun and ended, all that mattered of it, anyway. Saying good morning to her, being in her space, looking into her eyes was all I wanted and it was over in seconds.

Disturbed and stricken as I was, I suddenly felt alive. This was not, it could not be me. For decades I had kept people at arm's length. When I looked into this girl's eyes my defences failed me. I wanted her near. I could not, I must not, lose this vision, something carelessly swept overboard, never to be seen again. This is the face I want to remember when I can recall no other.

I must photograph her, capture her beauty for all time.

Her name was Joey. She considered my request in silence, gave me a long look.
"My camera wants to meet you" I said. "It will tell you how I see you".

We settled for 11.30 Sunday, two days later.

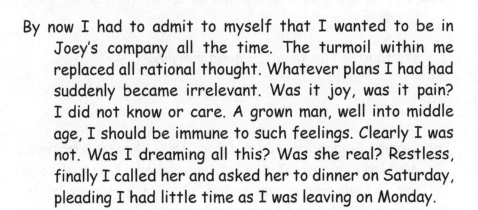

By now I had to admit to myself that I wanted to be in Joey's company all the time. The turmoil within me replaced all rational thought. Whatever plans I had had suddenly became irrelevant. Was it joy, was it pain? I did not know or care. A grown man, well into middle age, I should be immune to such feelings. Clearly I was not. Was I dreaming all this? Was she real? Restless, finally I called her and asked her to dinner on Saturday, pleading I had little time as I was leaving on Monday.

She stood her ground. She was busy Saturday.

"But", she said, "Sunday my time is yours. Lunch and dinner and in- between". And then she added: "I shall make myself beautiful for you".

Lunch………and dinner……and in-between………she will make herself beautiful…………

It took me a while to collect myself.

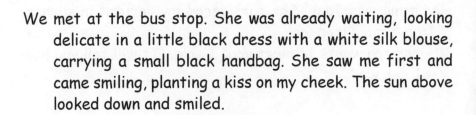

We met at the bus stop. She was already waiting, looking delicate in a little black dress with a white silk blouse, carrying a small black handbag. She saw me first and came smiling, planting a kiss on my cheek. The sun above looked down and smiled.

We walked down Hill Street like old friends. Our table at Raffles was booked for one o'clock. I reminded her of the camera but that could wait. We had lots of time. Did time really matter? A joyful thought.

I stopped to face her.

"How about ground rules for the day?! If you are bored or feel uncomfortable about anything, you will tell me right away, won't you?" "Right away", she smiled playfully and started walking.

Chijmes is a picturesque complex of restaurants and shops in what used to be a convent. It has a spacious inner courtyard. A church, no longer functioning as one, completes the rectangle. A good place to sit and talk.

We sat in the courtyard. A glorious morning with sun and cloud and a light breeze. The buildings around us shut out the world.

We talked at length. As I spoke she looked into my eyes and I felt she was looking into my soul. She listened with attention and did not interrupt. When she spoke what she said was interesting and charming.

Her perfect, oval face was framed by shiny black hair. The breeze picked up strands of her hair which brushed her cheeks and her lips. Her delicate eyebrows gave emphasis to the beauty of her eyes. Big, open eyes, everything revealed, nothing concealed. Her skin bore no blemish. An easy smile dwelt on her mouth.

I brought the camera out.

"Just talk to me, talk about anything," I said as I studied her face through the lens. The zoom brought her so close that my heart began to pound.

My invitation to photograph her was just that. No matter the impact her eyes had on me I had no ulterior motive. I just wanted her company. I was here to make a record as I had promised. A face to remember.

Yet, try as I might, it was hard to conceal the emotional impact of seeing her so close through the lens.

Joey had none of the affectations of pretty girls. She followed directions well. She was patient. The pictures promised to be good.

We walked to Raffles with a good feeling about each other. She was chatty and happy. There were no awkward silences. She loved the hotel courtyard and appreciated the architecture. Her joy became my joy also.

I was proud of my lunch companion as we entered the Tiffin Room at Raffles Hotel. Joey carried herself well and was at home in the fine, old-age restaurant where elegance is matched by the immaculate service provided by almost unseen waiters. Attendance at table is an art form, alive and well at Raffles. The curry buffet and desserts accompanying it are an experience for life.

As we ate I stole a look at her, only to find that she was looking at me. We looked into each other's eyes. Two strangers, just feeling comfortable with each other. Silence said it best.

"Your eyes are green", she said finally.
"Your eyes are dark, mysterious and beautiful."
"All Chinese eyes are the same."
"No,......not all......"

We were the only diners in our half of the restaurant, everyone else preferring to be seated near the buffet. We might as well have been alone.

"Try this" she said, offering me a tidbit on her fork.
"I can use my fork" I said but was flattered by her natural intimacy.
"Take it", she said and smiled when I put her fork in my mouth.

It was still early afternoon when, like a shadow, the thought of leaving the next day crossed my mind.
"Today is ours" I mused aloud.
"Tomorrow, a few hours." Joey said.

I looked at her across the table. Feelings and reason fought a savage battle within my chest. She must not know.

I thought of Omar Khayam:
   "...the bird of time
      has but a little way to fly,
         and lo, the bird is on the wing..."

I thought of Housman:
   "come to the stolen waters

> and leap the guarded pale
> and pull the flow'r in season
> before desire shall fail"

I had to say it.
I had to say it now.
"Joey"......
She looked up.
"Joey, I............I.................hmm..................I'm so glad you came....."
She looked at me for a while and nodded imperceptibly.

Lunch over, we spent time in the Raffles Hotel courtyard where the light is gentle and gives no unwanted shadows. It looked like daylight might last forever. I took pictures of her, even sang a song in atrocious Chinese. It made her laugh. I felt happy, alive.

"We must find a name for this spot" I suggested.
"Cinta Antara Dua Benua" she replied. "In Malay it means love between two islands."
Love............

Her cell phone rang three times during the afternoon.
"Your boyfriend?" I asked. Pretty girls are spoken for.
"Yes" she replied without hesitation.
"I don't want to cause you trouble."
"I'm honest with you in saying I have a boyfriend. I am also master of my affairs and I chose to spend today with you."
"How old are you, then?" She seemed so self-assured.

She whispered in my ear and gave an impish smile.

The world stopped turning. This self-assured, well brought up, confident, regal young lady was much, much younger than I had thought! She was mature beyond her years. And I? Whatever was I doing here?

I voiced my surprise. I complimented her on her poise and demeanour and uttered something incoherent, half-apologizing for my age. I felt protective. My unspoken feelings of romance, never intended, but oh, so beautiful, fled in disarray.

She was not fazed and told me just to be myself.

We walked with crowds, we walked alone. I let my eyes linger on her face, her eyes, her lips. I felt fulfilled just to be with her.

Was this the last hurrah of an aging man who wants to stay young? Was I a passenger on a train who forgot to get off at a station and was now carried helplessly along?

Did I not realize that I had left the land of youth long before, my passport had long since expired and would not be renewed? Were I to enter that land by stealth, all too soon my appearance, my words, my interests would unmask me as an intruder. And then the exile, which surely must follow, would be more painful, so much harder to bear than the recall, through mists of time, of days long gone.

Is a moment of bliss, a touch of eternity, worth all the pain? Why will a man climb the highest mountain and accept untold risk and agony for the privilege of a view that is his alone for a few fleeting minutes?

As the sun went down and light began to fade I wished, against all reason, that the day would last a little longer, that its inevitable end would, for once, arrive late. I knew that sadness was just around the corner.

We had dinner in fading light. Her cell phone rang again. I withdrew. She caught up with me, upset that I had left her. She had to go.

The spell was broken. We walked to the bus stop in silence and agreed to meet for breakfast at nine the next morning. She knew that I would have to leave for the airport soon after eleven.

A cloak of sadness enveloped me as her bus pulled away.

The next morning we met early. She was happy and eager to see me. I had the pictures ready. We were both pleased with the result.

Her beauty had been captured, my mission fulfilled.

Yet I had to admit that the camera revealed as much about myself as it had about Joey. The lens had worked in reverse fashion and threw into sharp focus a vulnerability I did not know I had. It dismissed the protective shell designed to keep others outside my perimeter. It made mockery of the arm's length attitude I had practised for so long.

The camera also reminded me that feelings of love and passion need not die with the passing of youth. Those moments of exaltation, so long gone, so briefly recaptured, will stay with me forever.

"Forever" itself is a concept open to interpretation. I had promised Joey that my pictures would preserve her beauty for all time. The joy this girl brought me will also last forever, but this is a personal scale of time which will end when my eyes are finally closed.

I must be honest with myself and admit that my rapture was merely an illusion. We had made a bargain and we both lived up to it. My unspoken, unintended feelings of passion and romance were of my own creation. Still, I held my feelings in check and did not intrude into her world. She noted my feelings graciously and spoke no critical word. I let my guard down and was captured in the web of my own imagination.

Yet when my life is lived and I recall my joys and sorrows, my day with Joey will be a treasured memory. It taught me that sometimes it is better to be the captive of an illusion than to stay in command of reality. The salaries of reality are soon spent. The joys of illusion may never die.

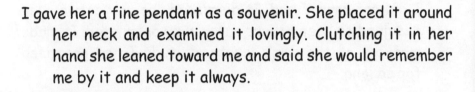

I gave her a fine pendant as a souvenir. She placed it around her neck and examined it lovingly. Clutching it in her hand she leaned toward me and said she would remember me by it and keep it always.

I told her about my sadness. I thanked her for bringing me the gift of youth for a few hours. In the autumn of my days she brought me a day of summer.

"Your sun is still rising, mine is going down."

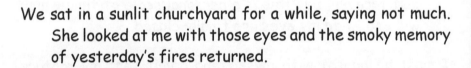

We sat in a sunlit churchyard for a while, saying not much. She looked at me with those eyes and the smoky memory of yesterday's fires returned.

Then it was time to go.
I held her close for a moment.
"You go first" she said.
"No, I'll watch you go."
She walked away.
As my tears welled up, I lost sight of her.

When I wiped them away, she was gone.

HIDE AND SEEK

The phone rang. Call from Emergency.

"There's a child for you to see". The nurse's voice was detached, impersonal, perfunctory. Must be the end of her shift, I reasoned.

"What is it?"

"Something abdominal"

"What's her name?"

"Dunno…..she says you have seen her before……"

God, please don't let it be her!, I prayed on the way to the Emergency Room.

It happened some years before. It was lunchtime, the X-ray Department deserted as I walked through. On the viewing box a series of films was waiting for the radiologist's return. I stopped in my tracks. The films were those of a child with a large kidney tumour. There was no doubt: the tumour was malignant. I collared a passing colleague and gave him a teaching session on the X-ray features. He paled. He was shocked, speechless. It turned out he was the child's physician and a close friend of her parents. There was no time to waste. Back in those days the kidney had to be removed early, with other treatments to follow.

Just an hour before, the fit little girl had been playing soccer during the lunchtime break at school when she was bumped in the flank by a fellow player. Immediately she felt pain. She was brought to the hospital. The seemingly trivial trauma was enough to reveal the awful truth which had been lurking undetected.

One of the heaviest burdens of being a surgeon is the duty to inform the patient, this time the parents, of life-changing truths. I did that. They understood that an operation was necessary as soon as the child was ready for an anesthetic. Of course, I had to talk to the little girl also. I told her that she had a "bad kidney" which had been injured and was bleeding and that I would take good care of her. She took it in good part.

The operation was done without delay, yielding a large, typical childhood cancer in the kidney. There was no evidence of tumour spread elsewhere within the operative field. We breathed a sigh of relief but knew that the child needed chemotherapy and radiation treatment to follow. There was no guarantee of cure. I told the parents and the family doctor to put the best face on the situation. They, in turn, had decided amongst themselves that the child not be told the truth under any circumstances. They agreed that I could tell the child that it was indeed a "bad kidney" and that it needed further treatment. Everyone remained upbeat within the little girl's hearing and we wasted no time in arranging a consultation with the Cancer Clinic. In the months which ensued she spent time there and received the full quota of treatment measures appropriate to her condition at that time.

Needless to say, you cannot keep an intelligent child at the Cancer Clinic for long before she stumbles upon the diagnosis. In the course of chemotherapy, layer by layer, the truth was peeled back. She did not lack for love, she felt safe in the affectionate surrounds of the home environment. She made good recovery. The word "cancer" was not mentioned in conversation. She remained under supervision and her follow-up tests were favourable.

During this time I was in conflict with the family doctor over the issue of hiding the facts from our patient. My policy was simple: I always revealed the truth to my patients, knowing that if I played games with the facts they would not believe me when they most needed to believe. In the wake of these conversations the little girl finally learned the truth. She took it well. I did my best to instil a spirit of optimism in her and appeared upbeat and happy in her presence. As time passed, the continuing good test results helped underscore this buoyant state of mind. What kept me awake at night was the question of how this child might deal with occasional squalls of doubt that surely must have paid her visits. If they did, she volunteered no such information. She was busy again at school and returned to playing tennis, her favourite sport, and did well in competitions.

My heart sank as I reached the Emergency Department. There was my little patient with an abdominal mass the size of an advanced pregnancy, a solid mass, leaving no doubt. The stark truth revealed itself: cancer was widely spread throughout the abdomen. Clearly, nothing could be done. No reassurance, no surgery, no medication, indeed, no prayer would make the tumour go away. She was anxious, she was in pain. The family doctor caught up with me: "she must not know the truth!", he still urged, apparently in keeping with the parents' wishes.

She was admitted to hospital, into a private room with a "No visitors" sign on the door. The nurses were instructed not to engage in conversation about her condition within the vicinity of her door.

"Will you take care of me? ", she asked.

"As long as you live!", I countered, trying to sound confident but her question was tearing me apart.

And then she asked the question of my nightmares:

"Am I going to die?"

A part of me died at that moment. I sat on the bed and held her hand. I was searching for words, searching for composure, for this was perhaps the most important question of her young life. I must answer her. I must answer her now. I waited for the lump in my throat to clear, so I could speak.

"We are all going to die, every one of us. We don't know when it will be. Yes, you may die before me but I promise I shall take care of you".

I held her hand until she dropped off to sleep. I tiptoed out of the room, found some privacy and wept.

There are times when the truth is so horrible, so unspeakable, so final that we are not capable of dealing with it. We seek refuge in a mist of truths and half-truths, as much for ourselves as for those to whom we are trying to bring relief. In the swirling shadows born of our compassion we seek to mollify pain, subdue fear, hold the inevitable at bay, trying in vain to halt its inexorable march. Those who are believers may resort to prayer and grasp a crucifix. Perhaps the best gift I could bestow was to hold her hand, keep her company in the darkness now descending upon her. The trust she still had in me was now reduced to the touch of my hand.

A quarter century is not enough to heal this pain. I am wiping my tears even as I write.

SPRING CLEANING

SPRING CLEANING

Word that I, a newcomer to the area, was afraid of snakes reached the ear of my old colleague Hugh Campbell-Brown. And so he came to telephone me on a sunny April afternoon.

"I thought you might like to help me clean out a den of rattlesnakes", he said, as casually as he might ask me to join him in a game of golf. His family had for long decades owned a piece of land which he was donating to the government to be used as a park. The wilderness and the climate made it an ideal habitat for the Pacific rattlesnake, native to our region.

All my life I faced my fears head-on and this was an invitation I could not refuse.

"Just what do I wear for this outing?", I asked. I imagined a suit of armour and steel boots as minimal safety equipment.

"Come as you are", he said, matter-of-factly.

"What about boots?"

"Sneakers will do".

By now I was committed but hardly reassured by his casual tone. Still, I turned up, wearing slacks and sneakers as he advised. We parked the car off the highway and walked up a cutting in the hillside.

"Get yourself a stick to use", he said as we climbed the hill.

I soon found a broken tree branch, about eight feet long which I knew would guarantee distance between me and the snakes.

"That will never do", said the old man. "Dry wood breaks too easily".

He pulled a knife from his pocket and cut a slender, four-foot branch from a Saskatoon berry bush. He swished it in the air, hit the ground with it and handed it to me.

"What you need is a switch, it is flexible and will not break on rocky ground".

He explained that the snakes coil up in the den in winter, keeping each other warm. The dens have a rocky overhang at the entrance with a flat piece of ground at the front which he called "sun parlours". The rattlesnake cannot control its body temperature. In winter the entire colony seeks refuge deep in the den, in a state of hibernation. When Spring arrives the snakes soak up the sun on the "sun parlours", still keeping close to each other, regaining their vigour and vitality. Then they venture out in search of food.

"You have to be careful, they are hard to see, their colour melts into the background." I noticed his pace had slowed, he looked right and left as we proceeded. Clearly, we were approaching the den.

At length he stopped, signalled with his hand for me to stop also. He motioned to me to come around and pointed with his hand. Before us lay the den, filled with snakes, like coils of spaghetti in a bowl. They seemed oblivious of our presence, save for one which looked straight at us, the front part of its body in an S-shape, ready to strike if we got nearer. We were only about five feet from the den.

"The rattlesnake can strike only a third of its body length, so you are quite safe", he said to comfort me. "Besides, it is a slow mover"

I gripped the thin Saskatoon branch in my hand firmly. It felt like an inadequate weapon against the thick, muscular rattlesnakes I was seeing. Still, none of them moved and the one on guard was motionless. I leaned over the den and took a photograph of my adventure.

"So, how do we clean out the den?", I asked with some trepidation, armed as I was with only a slender switch. Hugh was carrying a stake with a forked end and hooks on two sides.

"I shall hook them and throw them toward you and you go to work with your switch. The rattlesnake has a slender neck, its weakest point. You just hit them in the neck, that will immobilize, even kill them. Ready?"

Hugh had a sharp, weather-beaten face and exuberant eyebrows. As he stood at the mouth of the den with his stake he was the very image of Saint George about to slay the dragon. The dragon, in fact was a denful of snakes and it was I who had to do the killing.. Things happened quickly and there was no time for contemplation.

In no time I had four rattlesnakes by my feet, all suddenly wide awake and alert to the threat represented by this figure in sneakers who knew he had to kill them before they got a chance to strike him with their poison.

Circumstance is a great prompter in hazardous situations where instinct comes in to compensate for lack of experience. I had never before seen a rattlesnake, let alone hear its angry rattle. The urgent sound, together with the coiled, ready to strike posture leaves no doubt that it is a one-on-one confrontation, with possibly serious consequences. As it was, the face-off was four on one. Four snakes, all busy rattling, right by my feet. How I wished I had had an eight-foot stick to keep them at bay but as soon as I went to work I realized that it would have broken to pieces on the rocky ground and I would have been defenceless. Instead, the Saskatoon berry branch was ideal, it did not break or fray. In no time I stood over four immobilized, perhaps dead, rattlesnakes.

"Plenty more in here", said Hugh, sending snakes flying in my direction. By then I had lost all fear of rattlers and felt competent at killing them. One of the angriest was a baby snake, about a foot and a half long.

"Just as poisonous as the big ones", said Hugh.

With the den empty, or so it seemed, Hugh wandered off, down the hill, looking for any stray snakes for me to find and kill. After a few minutes he returned. Throwing his head sideways he said there was a rattler down there somewhere. I set off rather carefully, for this time the snake would not be served up on a platter. I walked gingerly but with no remnant of fear.

"Stop where you are!", shouted Hugh from above. "Take a good look around you".

I did as he told me. Right before me, no more than four feet away there was a rattler, ready to strike. As it happened, it did not rattle, so, but for Hugh's warning, it might have struck me. He did not rattle until I raised the switch to kill him, the last, defiant gesture in his life. I picked him up by the neck and took him to the den where all the snakes lay motionless. They must have been dead.

"That snake has good markings", remarked Hugh, "why don't you keep the skin?"

He told me to hold the snake with one hand and cut the skin on its belly from end to end, then peel it away sideways until I had it all. As soon as I cut the skin, the snake began to coil and rattle. Hugh told me calmly it was OK, they did it for some time after death. Still, it was unnerving to hold this muscular body heaving this way and that, especially when it began to rattle as well. Hugh just looked at me, amused, watching my baptism of fire. At length I had the skin removed. Underneath, the flesh was pink, "quite good eating", said my mentor, but frankly I had no appetite for flesh nourished by rodents. I took the skin home, stretched it on a cedar plank and gave it to the Silver Star school.

Hugh pressed the snake's jaws together. A pair of fangs sprang forth from the palate. "Hypodermic needles, carrying venom", said Hugh.

"The den will be populated again next year, so we'll come out again to do the same", Hugh told me as we made our way back to the car.

"But why kill all these snakes?", I asked.

"This land will become a park next year where families with children will go hiking. I do not want the children exposed to danger", Hugh replied solemnly.

As for me, although I do not enjoy killing, I had to admit I felt good about my experience. I faced one of my fears and I overcame it. Hugh is long gone now but I shall ever be grateful to him for his gift.

I still go hiking, looking for rattlesnakes but only to photograph them. They have a right to be there. I would rather have them around than be overrun by rodents.

Yet I retain the uneasy, guilty feeling that the snake I skinned was still alive, stunned but not dead, when I robbed him of his skin.

POSTMARK: TAKLAMAKAN

a letter
(If not delivered, treat as abandoned)

Indecision is like the desert of the Taklamakan, an endless, unfeeling, featureless landscape in which there is no reward for toil, where aimless miles carry no promise of shelter. I am lost in the desert of my indecision. It is best that I write to you while I can. I ask myself, how did I come to be here when only recently I set foot on your tropical island.

At first it was nothing more than a barely perceived thought of you between dreams. Soon the thought lingered just outside the window of my consciousness. Now, no longer do you visit my few wakeful moments. You are with me as

I awaken to the sounds of the night, my heartbeat and the heartbeat of the sea: the surf pounding the reef.

The truth is that since I first set eyes on you, you have occupied my world to the exclusion of everyone and everything else. I see people but they have no faces and they do not speak.

When I hear footsteps I tell myself they are yours. Any soft sound within my hearing becomes your gentle knock on my gate. I rush to open it and see only the lotus flower inviting my gaze before it closes its petals again.

Foolish of me, to think you might come to my door. You don't even know me.

We have not met. Yet, at my first sight of you I became enmeshed in this web of longing. Your classic island beauty would capture the heart of any man. Yet it was your reserve, your distant, unattainable look that has held me prisoner ever since. That far-off, lofty, unreachable Everest air about you, in quest of which men will risk all for not even a half-promise of success, has put down its challenge to me.

Each time I see you I uncover new mysteries in your face, a thousand questions to which I may never know the answer, a thousand and one nights of untold tales, mostly of my own invention. You have set my imagination free: a terrible curse if it comes to naught, yet a gift denied to so many men. I have painted vivid dreams of us together. I called out to you during those blessed nights on your island when the breeze had gone to rest, the air was warm and kind, the moon in full flight overhead and the crickets filling the darkness with song.

"We are such stuff as dreams are made on" mused Prospero, cast away on his island. I, too, am under a spell on this island, under your spell, let me admit it. You set the pulse of life beating within me. Far from admitting to myself that there may never be a "you and I", I revel in the fascination you brought me. I think of the multitudes of men who would forfeit time from their lives to experience again, or maybe even for the first time, the excitement you bring to me. A flight of fancy for sure but flying free on the wings of imagination I see the whole spectrum of life in one passing thought.

To think that you are blissfully ignorant of my feelings! Am I just a visitor to a gallery in love with a portrait? Time alone will tell. I always think of Time as a ruthless judge for it gives no reprieve and admits no appeal. Yet I recognize that Time will heal wounds and soothe pain. All I can hope is that its judgement will not be harsh.

Our relationship knows no words. I am acutely aware of your presence in those brief minutes when you are in my space. My gaze celebrates your beauty but not long enough for anyone to notice. Each time you walk away I feel the pain of parting.

You must know, you cannot fail to have noticed that I long for you. You see me sitting alone day after day. When I leave, you see me turn for a last look, not at anyone or anything else, but you.

And I, sitting alone, have noticed that you are conscious of my presence. Now and then you give me a fleeting smile, not in a brazen, challenging way but in the manner of a shy girl who is aware of a man's unspoken interest in her.

Those few moments when our glances meet are not measured in time. They have nothing to do with the passing of the hour or the flight of the sun from horizon to horizon. They bear no relation to the ebb and flow of tides or the phases of the moon. They are timeless, free from earthbound notions which would grudgingly call them a second, maybe two. In the realm of human feelings they are eternal. If there is any measure at all, it would dwell not on how long our eyes met but whether they met at all. That is when the window of the soul is opened for a glimpse inside. That is how heart meets heart.

But then, in that brief exchange of glances did our hearts truly meet? I fear not. Yet I tell myself they did because the thought is exciting and its denial leaves me feeling desolate.

Strange, we cannot even talk for you speak my language hardly at all. If I tried to talk to you at length, you would smile and shrug your shoulders in resignation as my words failed to reach your understanding.

You are like a rare tropical bird which has inadvertently wandered within my reach. Can I admire you? Yes! Touch you? No! Should I speak out? Better not...

I want to shower you with gifts I can afford to give you things you would cherish and want to keep. God knows

I could give you money which you could put to good use. Yet, money, ultimately the most versatile of gifts, carries a bad connotation. It would make my beautiful bird fly away.

Then I must remember that soon I shall leave your island Is it fair to kindle a flame in your heart when I know I shall not be around to tend it? Or should we just follow our hearts and give way to our feelings while we can? If we used our heads, not our hearts, would regret last forever? Reason is, at best, a caution sign, not a signpost to a destination.

Is there an answer?

Is there a choice?

This restlessness within me shows no sign of ending. It now occupies my day and invades my rest at night. Sleep is but a thin veneer penetrated by the slightest sound. The moment I wake, thoughts of you come flooding, crowding into my brain. I see you walking toward me with open arms, yet when you come close you just walk on by. I see your lips moving but hear no word. Then I see a deep chasm between us and run back and forth helplessly, looking for a path across, finding none. I try a hundred ways to help me sleep again but sleep eludes me. The more I realize that I want to sleep and cannot, the worse the torture becomes. Even if I am visited by sleep for a while, the restlessness, the longing will stay with me and by morning I shall be looking at the chasm that had stared at me in the night.

Where to turn? What to do? Do I now invade your life on the basis of your few fleeting smiles for me? Should I follow my passion for you no matter where it may lead?

The questions rush at me without mercy. I am no longer in the comforting lap of your island but in the searing desert of my indecision. How long before I come upon an oasis to soothe my pain, quench my thirst, make me whole again?

A hot desert wind has sprung up, sculpting the dunes into enticingly beautiful shapes. The sun casts weird shadows on the hills where grains of sand cling together, each one knowing that the winds of tomorrow will set them apart, carry them far away. New hills, new dunes, new shadows. Beautiful, but fleeting, like our time together.

The sun is hot, the wind relentless. Still no oasis in sight. The Taklamakan guards its secrets well. I am following a direction which has failed me before.

I have lost my way.

I doubt you and I will ever meet……

It is best that I entrust this letter to the desert wind. It has a better chance of reaching you that way. It will bear the postmark: TAKLAMAKAN………

"Taklamakan?"

…………it means you go in………

…………but you don't come out…!"

LA DAMA BLANCA - A PILGRIMAGE

LA DAMA BLANCA - A PILGRIMAGE

When news reached me that the Chilean Navy training ship *Esmeralda* was due in Vancouver I booked a flight there at once. My old school friend Michael Woodward met his end under torture aboard that vessel at the time of the Pinochet coup in 1973. We had lost contact after his last letter to me in 1954 at which time he entered a seminary in Santiago, Chile. Our lifestyles in pursuit of our separate careers were not conducive to writing long letters. When, in 1973, I saw his name in the obituary column of the old school magazine I did not connect it with the evil political transformation that had taken place over there. The fact is that as a young parish priest in an impoverished area of Valparaiso, he devoted his life to helping the cause of the poor parishioners among whom he lived. The kindness he extended to me when I was a poor scholarship student at our privileged boarding school in the west of England he continued to bestow on those around him, improving their lives and their prospects.

The CIA took a dim view of the democratically elected left wing socialist government of Chile under its president Allende and with the cooperation of the Chilean Navy staged a bloody *coup d'état*. Chile was a country populated by a majority of "have-nots" but ruled by a rich elite to whom the plight of the ordinary man was of no concern. Those who had tried to help the poor, including President Allende and, as it happened, Michael Woodward among many others, were labelled Communists or at least their fellow-travellers. The President's palace came under fire by aerial and artillery bombardment and he himself was killed in the early hours of the attack. Michael and thousands of others with him, were ferreted out, arrested and tortured. Many testimonials survive of the inhumane torture methods inflicted upon them.

Michael, it is known, was taken aboard the *Esmeralda* where he underwent interrogation. As described in a book (*Blood on the Esmeralda*, Cruzet, Downside Books, 2002, ISBN 1 898663 14 9), he was severely beaten and "when the answers to questions were not forthcoming or considered unsatisfactory......(the interrogators)......adopted the method of wrapping their fists in wet towels which would prevent their punches from leaving marks on the victim's body". Michael, a lean young man, with no body fat to protect him, obviously suffered internal bleeding. A doctor who saw him described his high pulse rate as "auricular fibrillation" when it was clearly shock, due to blood loss. An ambulance was called to take him to the Naval Hospital. Michael lay on the deck of the ship and died before the ambulance arrived or maybe on the way to the hospital. Officially, he died of "heart failure". I suppose the heart does fail when there is no blood left to pump around.

The place of death was designated as "on the public highway". The body was not released to the Church or the family and was reportedly buried in a mass grave along with others. In any case, the diocese was not about to challenge the new regime and washed its hands of Michael Woodward. According to some reports, a highway system was built over the burial site soon afterwards, leaving no chance of ever locating the body.

The Pinochet government steadfastly denied that Michael, or indeed anyone, underwent torture on the *Esmeralda*. High-ranking surviving naval officers, even after the departure of Pinochet, continued to deny the fact, despite emerging testimony of many survivors who had undergone interrogation and torture on the ship. I question whether the long arm of the CIA is sealing their lips even so long after the event. More recently, remember Abu Ghraib, "rendition flights" of prisoners and other well-known, notorious places of detention.

The facts that have filtered out through testimonies of courageous witnesses will continue to be denied by those who know all the answers but feel secure in their lies, protected as they are by forces beyond reach. Imagine if at the Nuremberg trials the same protection had been proffered to the accused Nazis!

Little wonder that at whatever ports the *Esmeralda* visits during its annual training cruise, it is met by protesters, many of whom had been tortured aboard and who know the secrets of the ship all too well. For them, the persistent denials and even the very presence of the ship keep re-opening old wounds that will not heal. These thoughts were passing through my head when, crossing the Burrard Inlet on the *Sea-bus* I first sighted the tall masts of the *Esmeralda* tied up at Lonsdale Quay. An emotion gripped me, I felt I would rather be anywhere else but not there, but we arrived at the terminal all too soon. Within minutes I stood looking at the stern of this undeniably beautiful sailing ship, The White Lady, *La Dama Blanca*, as it is known. The flag of Chile waved lazily in the breeze. Aboard, young sailors in immaculate white naval uniforms went about their Sunday duties and a contingent of them sauntered past me to some destination on shore. They were all young, bronzed by the sea winds, too young to have been alive when the event, the object of my pilgrimage took place.

A loud group of protesters stood on the quay, carrying signs demonstrating against the presence of the ship, showing photographs of Michael Woodward, taking turns at the megaphone, uttering laments, all joined in spirited chorus from time to time.

In the face of all this the young crew members conducted themselves impeccably, no doubt primed for such encounters by prior instruction. Still, the day being exceptionally beautiful with a canopy of blue skies and the gentlest of breezes, I knew it was marred for them by the sights and sounds of conflict, not the first since their journey began along the western coast of North America. Behind the tight lips and unfailing good manners you could discern the pain they felt when, instead of being welcomed by smiles and handshakes on this, for them, a proud day to show off their ship, they were made to remember the ominous shadow that follows in its wake.

At long last, we were allowed on board. It was eerie to think what Michael's thoughts may have been walking the same boards as he stepped on the ship. I am sure he did not dwell on the glittering shine of the brass or the uncannily tidy way the ropes were stored. He must have anticipated the trials that would follow and wondered if he was equal to living up to his principles, no matter what torments lay in store. By then he knew that this was his lot in life, yet I doubt he realized how little of his life remained.

Above me, the four masts, 48 metres tall, as I learned, stood like exclamation marks with the rigging a maze of lines the sailors had to master in bad weather and fair, no matter how the ship pitched and rolled at the dictates of the sea. I always admired anyone who would do his seaman's work on so precarious a perch, knowing I would not be able to rise to such a calling. The ship's bell, engraved with the name *Esmeralda*, was polished to a fine shine. Behind it the wheel, merely symbolic now, spoke of the old days of tall ships.

The crew, some of whom spoke English and others who merely tried, conducted around those who like to travel in groups. I preferred to find my own way, keep my own company, think my own thoughts. There was no access below decks, where the interrogation took place. I heard a one-eyed old man ask a sailor about "the man who was murdered". He was met by the curt response "we don't talk about that!" For the first time the ship had female midshipmen on board, well turned out and ready to smile, softening the impact of an otherwise painful experience.

Finally I came across a stretch of deck, where, according to the book about Michael, he must have spent his last minutes. This was the place I was seeking. An immense sadness came over me. I tried to say a prayer but realized it was futile from a man who has not prayed in years. I just looked at the planks on the floor, the gangways, the rail and knew that the tears that welled up in my eyes would be accepted by Michael as a sufficient parting gesture. Suddenly I felt the closure that comes when you turn the last page of a gripping narrative and realize that the time has come to detach yourself from it and get on with your own life. I walked off the ship content that I had honoured my old friend, sad about all the years I did not write to him while I was busy with my own existence, humble in the knowledge that my grief at his fate was no more than the grief of millions, all just visitors to our world, who, after grieving for others, will leave others to grieve for them when their time runs out.

The protesters were still vocal as I left the quay. I asked myself, what has been achieved by protest over all the years. Waves crashing on rocks, as they have done through the ages.

There and then I resolved to do something different, something to break the deadlock, the futile interplay of protest and denial. This could go on until there was no one left to remember. The victims would fade away unrecognized, the medals of the admirals buried with them until they, too, would sink into oblivion. As for Michael Woodward, I was saddened by an annotation on the cover of the book about him which read "readers can judge for themselves whether Father Woodward was a fool or a martyr in Christ, or possibly both". He would not name his friends to his interrogators. For that he died. Was it not said that "greater love has no man than one who gives his life for his friends"?

I wrote a letter to the President of Chile, asking him to consider a proposal, a bold, conciliatory gesture which "would once again reassure the world of the good will of the people of Chile, restore high regard for the Chilean Navy abroad, extinguish further protest and recognize the selfless dedication of Michael Woodward to the disadvantaged people of Valparaiso, home port of the *Esmeralda*"..

The proposal: re-name the *Esmeralda* to the **Miguel Woodward**.

Cynical as I am, I did not expect an acknowledgement. It came as a surprise, therefore, when I received a letter stating that my proposal has been placed before President Sebastian Pinera for consideration. But that was years ago............

By now more years have passed. The absence of further word from the President's office would seem to indicate that no action will be taken in response to my letter. Those responsible for Michael's death, if indeed they are still alive, rest easier with each passing year. The shadows of the Pinochet years still persist, the sinister power of politics will ensure that the protesters will either give up in frustration or die, while the guilty look out of their hiding places with a "you can't catch me!" smirk.

I doubt that they will suffer any pangs of conscience over that. The sad conclusion is that "bringing democracy to the world" can accommodate such hypocrisy.

In about 2013 two men were given short prison sentences in Chile for their part in Michael Woodward's murder.

HOW DO YOU SAY IT IN GERMAN?

Tom and Lilian were a refined couple, not given to excesses. They lived a life of quiet, comfortable elegance and in their house you could always expect good conversation, hearty laughter and exceptionally good food. What happened on this occasion, therefore, caught me off guard, not ready to mount a response.

We were well into a fine consommé when Lilian touched me on the arm and with an earnestness signalling deep and troubling concern she said: "Nicholas, do you know where the word 'fuck' comes from?" She was the last person I could think of who would pose a question like that, let alone pronounce the word without at least a perfunctory attempt at muting it to limit its impact. As it happened, I had a pretty good idea of its Anglo-Saxon origin, but I at once recognized that Lilian had found an etymological gem and therefore chose to let her advance her theory.

"No, as a matter of fact I don't", I replied.

The light of triumph sparkled in her eyes and she was ready to put the missing piece of the jigsaw into place.

"In the Middle Ages they placed fornicators into the stocks to expose them to public shame and ridicule. The stocks bore a sign proclaiming the reason why they were so punished. The sign read "F.U.C.K.", an acronym: 'For Unlawful Carnal Knowledge'". She leaned back in her chair, satisfied to have imparted this vital piece of information.

The consommé was delicious and I elected not to challenge her thesis. Instead, I just uttered "remarkable!", in a tone that implied my indebtedness for this piece of knowledge. Deep down I felt that there had to be a more earthy explanation.

As it happened, my friend John owned the definitive tome to be consulted in such matters. Two hefty volumes, printed on the thinnest paper with letters so small you had to use a magnifying glass to read the contents, were called *The Compact Edition of the Oxford English Dictionary*. "Compact", true British understatement. Clearly, this was the seminal source that would yield the answer I was seeking. Before I would delve into the book I casually asked John the question that Lilian had put to me. John shrugged his shoulder and was not forthcoming with an answer but he blew the dust off the volumes and with the magnifying glass in hand set off in search of the word "fuck" in the enormous book.

After a while of anxious waiting he turned to me and gravely informed me that in *The Compact Edition of the Oxford English Dictionary* such a word did not exist.

"But John, it is not possible! "Fuck" is one of the most pivotal words in the English language. It is a noun, a verb, an adjective when modified, an exclamation to express admiration, grief, joy, surprise, disappointment or just a word to use when nothing else will do! It is a hyphen to link other words together. Segments of the population could not communicate without it. What the joker is to a pack of cards "fuck" is to spoken English. For versatility as a word it has no equal!"

John, anxious to help, appeared crestfallen at my frustrated outburst. Still, he had done what he had to and could go no further. We sat in silence, contemplating the failure of our quest. I felt like sending a letter to the Oxford University Press in protest.

Then an idea came to me. A good Anglo-Saxon word like that would have its equivalent in German. That is where our friend Karl came into the picture.

The telephone kept ringing. I was about to give up when at length Karl picked it up. Now I should say that I had only a superficial acquaintance with Karl and beyond knowing of his Swiss-German origins and some family ties to John we had little common ground.

"Hello, Karl, this is Nicholas" I said in a hearty tone.

It was clear that Karl did not at first know who this Nicholas was. "How are you?", he stalled, waiting for clues to reveal my identity.

"I'm fine Karl, how about you? I have not seen you since the wedding." He was now close to identifying me, judging by the rising confidence in his tone.

"I'm fine too! It's good to hear we are both fine!" he allowed, clearly on the right track now, aware that we had just about exhausted our common interests. It was time to move in for the kill.

"Karl, how do you say 'fuck' in German?"

Silence followed but I knew the line was not dead.

"Karl............are you there?" I suspected his wife may be in the room.

"I I knew it once but I have forgotten it ", he ventured.

"You mean that a red-blooded Swiss boy does not know the answer? How can one forget such a thing?". Obviously he was not alone.

"Let me call you back ", he said in a hushed voice. Clearly, he was heading for the basement or some other room in the house where he could not be heard.

The telephone rang.

"Nicholas", spoke a voice in the merest whisper, "is that you?"

"Yes it is" I whispered conspiratorially. "Is that you, Karl?"

"It is me", he stated, almost out of breath.

Time to get to the point while the going was good.

"So how do you say 'fuck' in German?"

The answer was barely audible, safely beyond his wife's hearing.

"Fick", he said, "ficken". He added: "it is a bad word, Nicholas".

"You are a brave man, Karl! You have done great service to the English language. Danke schön and auf wiedersehen!"

A Swiss boy, who had long forgotten it nevertheless came forth with the answer denied to me by the massive *Compact Edition of the Oxford English Dictionary*. Karl provided a clue, but maybe not the etymological answer.

I still don't know where the word "fuck" comes from.

Did the English language derive it from the German? Or did it spill from the stocks into German?

Was Lilian right after all?

STRIKE THE IRON

That hearing a certain language spoken could evoke painful memories was not apparent to me until a car trip in Italy many years ago. My companion and I shared the vehicle with two Italians who, as is their custom, conversed *molto vivace* in their own language. Now and then one of them turned and addressed me in English. I complimented him on his command of the English language. He fell silent for a moment, then with a face contorted with pain he responded: "..........signor, it is *doloroso* for me to hear English spoken!" It turned out that during the war he had been held prisoner by British troops, an experience that forever after coloured his views of anything English.

It is time for me to make an admission that I have kept to myself for decades. I lived through the same war, witnessed and felt its horrors at close quarters, saw the once beautiful city of my birth reduced to total ruin, friends taken away never to be seen again, gunfire beating a staccato message of fear in the night, followed by the silence of death, vivacious men and graceful women reduced to lifeless corpses. The occupiers saw to it that they left behind nothing but "scorched earth", as they called it. All this to the sound of harsh commands spoken in a language that showed that these terrible deeds were enacted by fellow humans who, for reasons of their own, had shed their humanity. Forget it I cannot, not if I live a hundred years! After all these years the sound of their language still conjures up images I do not wish to recall.

It is not in me to harbour dislike or hatred. Invariably I try to find a link to put an end to differences, to find common ground. It bothers me that the sound of that language remains an unfiled page in my memory, a needless barrier between myself and the blameless individuals who speak it. Long ago I resolved to put the matter behind me. Why cling to a memory I want to forget? The trouble is, the memory clings to me. Either way, it must be shed.

When I was a boy I found myself afraid in a dark cemetery one night. I convinced myself that the way to cast off that fear was to walk in the cemetery night after night and stare it in the face. It worked. I knew that to overcome this prejudice, for what else is it, I must get to know someone who speaks this language, someone to put my resentment to rest. Easy, I decided, but there was always tomorrow, or the next day, never now. I longed for the opportunity but was tardy in creating it.

Opportunity found me. While having breakfast at a resort in Indonesia I heard the language spoken at a table behind me. The same old reaction visited me. I tensed up. I chided myself: I could hardly be further away from my wartime experience. Let the matter rest. But then, this was my chance to come face to face with something I had been avoiding all these years.

Then I saw the speaker. He was an elderly man with white hair and the kindest countenance. This face could not belong to an evil man! I felt an instant kinship. He was about my age. He must have been through all the fear, the cold, the hunger I suffered during the war.

He was my man.

There and then I decided that the next day I shall create an opportunity to make his acquaintance. Instantly I felt better, knowing I would at long last be relieved of my burden. Tomorrow it is!

Before I went to breakfast the next morning I enquired at the front desk as to his name and where he was staying.

"Sorry, Sir, he left on the midnight plane last night......"

CLOUDS FOR COMPANY

A daydream is an ambition put off till tomorrow. I had a daydream. Tomorrow came and went. Other tomorrows followed in long succession. Then it happened. My daydream came true. My elation knew no bounds. Alas, I woke to find it was just a dream.

The very next day I found myself driving by the airport, craning my neck at the airplanes basking in the sun. I had been doing just that for as long as I could remember. I made an illegal U-turn and my daydream ceased to be that.

I signed up for flying lessons. The man at the desk sized me up with an amused smile. He saw, he knew, how much this meant to a man of fifty-six. He assigned me to Dave, my first flight instructor.

"You're flying tomorrow, be here at nine"

An immense weight was lifted from my shoulders as I walked out. All my life I wanted to fly. All my life I found reasons not to do it just yet.

I now realized that a new chapter of my life would start at nine in the morning.

A dream will not come true if you wait a hundred years. It will appear now and then, tease and entice you, only to vanish until it returns again.

It is like that with all the things you wish for in life. You can sit in your armchair yearning for this or that but the object of your desire will not come to you. The world goes on outside,

No one knows or cares about your longing. In the end only you will know that time has passed you by.

Moments before that illegal U-turn I realized that if I had a dream I had to pursue it on my own, there and then, without looking back.

Released from the shackles of a long-unfulfilled ambition, at long last I felt free.

When I was a child I had a little canvas airplane which I used to launch high in the air with a catapult. I marvelled at the grace with which it rode the air and its habit of settling gently back on the ground. I watched the small airplanes of the day with wonder, admiring the courage of those flying them. My father had a friend who was a pilot. To me he was a man apart, one who conquered gravity, someone who could fly higher than the highest mountain I had ever seen. I longed to be like him.

David waited for me smiling as I arrived at nine the next morning. His manner was friendly and relaxed. I was full of anticipation and not at all apprehensive. We talked about the principles of flight and the function of the aircraft controls for a while, then walked out to a handsome, shiny Cessna 152 GQDJ (Golf Quebec Delta Juliet). I knew that the pilot always took the left-hand seat, so, once our pre-flight inspection was done, I crossed over to the right. "You are the pilot, you take the left-hand seat" said David, strapping himself in on the right. My world was changing fast.

Once in the cockpit with the doors closed, I felt I was in another world. The discipline that is flying asserts itself there and then.

A detailed, painstaking checklist begins, to ensure that all is well with the airplane before we leave the ground. As time went on, I came to view this procedure with a degree of fondness. It helped shut out the world and made me concentrate on the job in hand. It told me that the freedom of the air comes at a price of discipline. It gave me a sense of reassurance as the checks followed one another.

Then at last we were ready to fly. "We'll do some taxiing first", Dave said as he pointed in the direction of the taxiway. "Just a little power, steer gently with the rudder pedals".

I set off along the taxiway, the airplane reeling from side to side, like a drunk on his way home on a Saturday night. "Gently with the feet", said David, seemingly not at all concerned. I quickly learned that the rudder responds to the slightest pressure on the pedals and wondered how I would do during the takeoff run.

At length we were on the runway. A magical moment, to see the clear path ahead, a gateway to the sky, a hearth to return to when our flight is over. "Keep your left hand on the column and your feet on the pedals to feel what I'm doing", said Dave. The propeller was turning in earnest now, the engine roaring with impatience. I felt David take his feet off the brakes and apply full power. We began to accelerate down the runway. The centre line markings were quickly swallowed up as we raced along and before I knew it we were in the air. "Keep her straight with the rudder" said Dave but as soon as I tried we veered off wildly.

During the next hour I learned how responsive the controls were to the slightest deflection. David made the airplane dance in the air and she responded like a young girl dancing with light feet. I managed to fly straight and level, which was the objective that day. I made the radio calls I was instructed to do but had trouble deciphering the responses amid all the static. The hour passed quickly and soon we were back on the ground.

Flying the airplane is just the tip of the iceberg. Much more time has to be devoted to reading about engines, navigation and the thousand other things which are relevant to being a competent pilot and all of which must be retained in memory for the tests required. It was all fascinating. The more I read, the more I wanted to read. There were books on the theory of flight, on honing your basic skills, competent handling of emergencies and on how to stay out of trouble in the sky. They all helped my confidence grow.

Meanwhile I continued my lessons in the air. On only the fifth occasion David briefed me on spins. An aircraft out of control may go into a spin and it was important to learn how to get out of it. I must say that when David first demonstrated the spin I was totally disorientated, not knowing which way was up. Nevertheless, the spin was exciting and, once you knew what to expect, quite logical. I began to enjoy it, though not without some initial apprehensions.

Certain principles had to be learned. "Flying is not inherently dangerous but it is very unforgiving of inattention." "Keep cool but don't freeze", went another. I was told never to fly into cloud or fog, for, if I did so, flying without visual reference to the horizon would give me a life expectancy of 178 seconds. It seemed a good reason to avoid clouds.

To illustrate the latter point, one day when the air was smooth, David told me to set up the airplane in straight and level flight. If you do this correctly, you can take your hands off the controls and the airplane will fly itself. We were heading south, with the sun high in the sky. "Close your eyes now" said David, "see if you can keep her straight and level". I closed my eyes and held the column gently in my left hand. My feet rested on the rudder pedals without giving pressure. Even with my eyes closed I sensed that we had not changed our position relative to the sun. "How does it feel?" asked David. "Just fine", I replied, feeling somewhat smug about the way I was handling the aircraft. Flying is simple, I thought, I could do it with my eyes closed! "Good", said David, "just keep her straight and level." The sun then seemed to move a bit, judging by the source of light perceived through my closed eyes. I gave slight rudder to correct my direction. I must have overdone it, I thought, so I gave opposite rudder.

The sun seemed to have moved away but my body told me we were flying straight and level. "How does it feel now?" asked David a few seconds later.

"Straight and level!" I replied, feeling proud of myself.

"All right", said David, "you can open your eyes now."

I did.

We were in a spiral dive.

Flying with an instructor puts you through all the elements of flying with the reassuring presence of a seasoned pilot beside you. While you practise take-offs, landings, spins, stalls and the sundry emergencies that may confront you, your instructor quietly records your progress. Every student pilot lives for the first solo flight, the definitive transition from student to pilot. It is the instructor who decides when you are ready. My instructor at that time was Geoff, an old wartime pilot and, later, instructor, who had been flying for fifty years. I told myself this was a good omen because flying for fifty years and being still around must mean that you are very lucky or very good.

Geoff turned out to be the latter. He could make the engine purr like a kitten. He taught me how to gain altitude quickly by using the flaps. He taught me that, if my instruments failed 20 degrees of flap and 2000 r.p.m. would give me a safe airspeed of 70 knots. He taught me to come in to land fairly steeply to ensure a safe landing. "A good landing begins before you join the circuit, so pay attention early", he would say. "When you state your altitude on the radio, fly that altitude precisely, do not wander 50 feet above or below". He impressed on us: "When you are flying, you must know three things at all times: how fast you're going, what direction you're going and how far your arse is from the ground!"

Late one afternoon we were flying circuits. This is a busy exercise. As soon as you are airborne you must climb to circuit height at 1000 feet above the airfield, then go around in a rectangle, making all the proper radio calls and pre-landing checks and maintain the correct airspeed during your landing approach. Once you have landed, you keep rolling, raise the flaps, keep the aircraft in the centre of the runway, push in the throttle and the carburetor heat in the same hand motion, check oil temperature and pressure as you accelerate, then look at the airspeed dial. When you reach flying speed you pull the aircraft off the runway with your left hand on the column.

I was flying well that day and when, after about five circuits Geoff asked me to do yet another, I knew something was up. As we landed, he asked me to pull off to the ramp but keep the engine going. He handed me a piece of paper – my passport for solo flight. "Take her up for one circuit", he said, opening the door. The wind from the propeller swept his white hair as he walked away. My daughter Kati stood by the ramp. I wonder what she was thinking.

My joy was immense. The dream I had the night before the U- turn had come true! I was about to fly solo for the first time, accelerate down the runway all by myself, climb, turn, fly and land. Forever after I could tell myself that I had flown alone.

Discipline took over. This was no time for celebration. I must live up to the confidence placed in me, do what I was taught and not let Geoff down. I was eager to fly. The airplane stayed on the center line during the takeoff run. With only one body in the airplane, we got airborne quickly. The climb in the late afternoon air was smooth.

I touched the right hand seat – it was unoccupied. I was alone in the sky, just as I had dreamt. The scenery below was the same I had seen many times, yet it was different now: I was "captain" of the 152!

After all the handshakes on landing I walked back to the airplane. I patted it fondly. The engine cowling was warm. Leaning on it I gazed up at the sky. I knew my heart would stay up there forever.

Before you take your final tests for a private pilot licence you must complete a solo cross-country flight of some 250 miles and land at two distant airports on the way. You must prepare a detailed flight plan, taking into account your intended altitudes and airspeeds, as influenced by prevailing wind directions and velocities at the altitudes flown. The discipline of flying truly comes into play in the preparation. Naturally, you must follow a predetermined course and monitor your track by constant reference to the map and the unfolding topography. Given the beauty of the mountains, valleys and lakes of the Okanagan area of British Columbia this is also a sightseeing flight like no other. When you complete it, you can truly say that you have flown solo.

My course from my home airport in Vernon took me to Kamloops at the junction of the North and South Thompson Rivers. I had to land in Kamloops. From there south, to Princeton, then land in Penticton before heading home. The weather report promised scattered cloud and moderate winds from the west, nothing threatening.

On the way to Kamloops I found the air quite rough, giving me an uncomfortable ride. I consoled myself with a chocolate bar, laid on the floor in front of the passenger seat, only to find it was melted by the hot air vent. I landed in a brisk headwind without incident. By the time I was ready to take off, the wind, equally brisk, had changed direction 180 degrees.

I settled in the seat, map on my lap, to enjoy the long leg of the flight to Penticton. I had my flight log handy, to give me the planned heading. Some 20 minutes later I found myself well east of where I should have been. There was only one explanation: the wind from the west was much stronger than forecast. Knowing the points I had to fly over, I altered heading and was soon on course. Well ahead, somewhere over Princeton, I noticed ominous clouds. However, all was well and I was enjoying myself. I made sure I followed the intended track on the map. The ground under me was a high plateau but there was adequate clearance between it and the clouds.

All hell broke loose when I reached Pennask Lake. At first I found myself in a snowstorm with large snowflakes coming at my windshield like flak. Luckily, this did not last long. What followed was much worse.

Suddenly the air became very rough. The airplane dropped without warning, then was tossed upwards violently at the mercy of the wind. A wingtip would drop, requiring heavy pressure on the opposite rudder. I felt like a feather in a vacuum cleaner. This turmoil repeated itself in haphazard order for several minutes. My attention was totally occupied with trying to keep the wings level from one second to the next. Golf Quebec Delta Juliet stood up bravely to all this. Gusts of wind came from nowhere, like sucker punches in a boxing ring. With each blow I felt my body straining against the seat and shoulder harness. There seemed no end to the ferocity of the buffeting. Still, we were flying, Delta Juliet doing everything I asked of her.

Where we were, Pennask Mountain rises to over 6500 feet and neighbouring Mount Gottfriedsen to almost that. I was flying at 7500 feet and I wished I had more clearance. I kept looking at the Pennask Lake below, still frozen over, as a possible landing site in case I got into trouble.

I looked in the direction of Princeton. All the dark clouds and bad weather seemed to come from that direction and it was obviously snowing there. Reason dictated that I cut out the Princeton overflight. I decided to call the Penticton tower and advise them of the change in my flight plan.

Still on the Kamloops frequency, I reached for the radio to switch to Penticton. My hand sweated so badly that I could not turn the knob. We were still being tossed about constantly and I had not planned for a handkerchief or towel as essential equipment for flying. I rubbed my hand on my flying jacket, my trousers, the empty seat next to me, even my hair, but nothing worked. I decided to circle at this altitude until I made my radio call. I breathed on the palm of my hand, trying to dry it, when I could safely take it from the throttle. I did not forget to fly at "maneuvering speed" to reduce stress on the aircraft.

At long last the knobs obeyed my sweating hand. I called Penticton and told the controller I was experiencing severe turbulence over Pennask Lake, was cancelling the Princeton overflight and would head straight for Penticton.

"What damage is there to your aircraft?" asked the controller.

"No damage", I replied.

"In that case you are not experiencing severe turbulence. Report when you reach Summerland".

Not severe turbulence, he said! What about the agony I am going through, the desperate fight just to keep level, my racing heartbeat, the danger? Is my sweaty palm not evidence enough? Easy for you, sitting in your chair! Bet you anything the floor is not moving beneath you!

The best way to Penticton from my present position was to fly the valley between Pennask and Gottfriedsen down to the Okanagan Lake, then follow the lake to Penticton.

As luck would have it, that valley was occupied by a cloud, gleaming in the sun. I reasoned that if the wind is as strong as it is, the cloud would move. Meanwhile, there was nothing to do but circle in the rough air.

My reasoning paid off. Soon I saw a sliver of Okanagan Lake and I headed for it. Descending over the Brenda Mine I marvelled at the velocity of the wind blowing us sideways. Soon I was over the lake, low enough to avoid the rough air tumbling from the ridges to the west. I made a not-too-distinguished landing in Penticton in full view of a passenger aircraft waiting to depart. My legs felt wobbly but the blessed ground was still. There was no welcoming committee, no band, no speeches.

A severe crosswind was blowing as I began my take-off run for the flight to Vernon. Despite full rudder, I could not keep the aircraft straight and had to abort the take-off. I was cleared for another try. It worked. I flew high, in a straight line in pretty rough air. As I made my final approach into Vernon airport the air was unusually turbulent, so I decided to go around and try again. "Pretty rough air!", I called on the radio. Richard, the owner of the flight school, replied: "It's calm at the treeline". I looked at the ridge to the south and could not make out any treeline. Still, I was busy with another attempt at landing. This time it worked.

Richard came to meet me as I stepped out. "Some treeline!" I scolded him "Where is it?"

"Oh, I meant the trees at the end of the runway", he grinned.

There was no wind at ground level.

I have to admit it took me a couple of days to regain my composure after the harrowing cross-country flight. My next flight was also rough and I felt very uncomfortable. It was a while before I came to accept turbulence as a fact of flying and while I do not like it, it does not bother me now.

When I look back now I see the incident in a positive light. Very early in my flying career it taught me that the weather was a powerful force, to be treated with respect. "As flies to wanton boys are we to the gods", mused Shakespeare. But for the anachronism, one might think he, too, encountered bad weather while flying over Pennask Lake.

I also learned that, even as a low-hour pilot, I could deal with a severe challenge. Sure I was afraid and as removed from human help as could be, but I did not cease to function.

The way the little Cessna 152 stood up to the challenge made me love her even more.

As my flight test drew near I spent my air time correcting mistakes to which I was prone: losing altitude in steep turns, incorrect approach to emergency landings and landing to the left side of the runway centre line. The time had also arrived for practising solo spins. The spin is not the favourite maneuver of most students but I knew I had to master it. To put the aircraft in a spin you must slow it down with the engine at idle. You watch the airspeed drop, keeping the wings level. At the point of the stall you pull the column into your stomach. This ensures that the aircraft is properly stalled, that is, it has not enough speed to maintain flight. As the stall is achieved, you give full rudder on the side to which you want to spin.

The result is instantaneous. Sudden rotation occurs and all visual reference disappears for a moment. You go over on your shoulder, as it were. The next thing you see is the ground rotating fast with the aircraft's nose pointing vertically at it. By pressing the opposite rudder, rotation quickly stops. Allowing the column to move forward will break the stall and let the aircraft pick up flying speed while you level out and can apply power once more.

I had practised spins with my instructor but this time the right- hand seat was unoccupied. Needless to say, I was apprehensive as only a fool would willingly exchange perfectly normal level flight for a violent nose-down spin. I thought of the day I made the U-turn and then I knew I had to follow the course I had set for myself.

Before I throttled back I went through the sequence in my head. I knew what to do: it was time to do it. The aircraft slowed obediently. At the point of stall I gave left rudder. When the ground began to rotate I stopped the rotation, broke the stall and levelled out. The experience was so pleasant that I broke into loud song while climbing back to altitude for another spin. I practised spins to both sides and for the rest of the day I walked a few inches off the ground. From then on I did spins just for fun.

Spin entry - rotation is instantaneous

Nicholas Rety

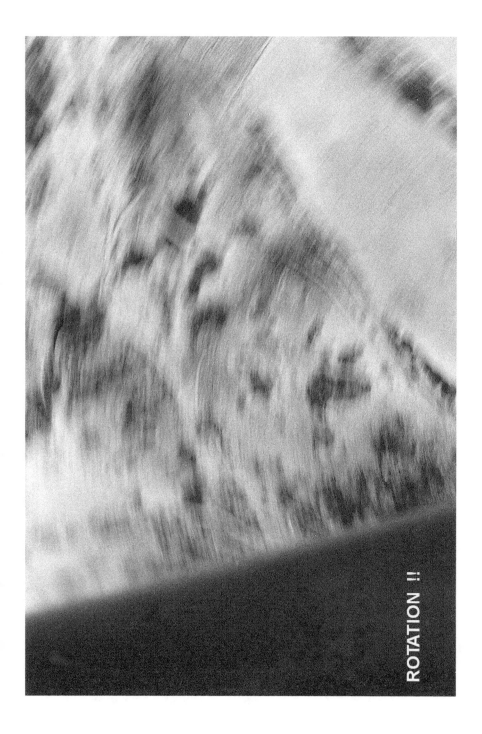

One bad habit I developed was to raise the nose of the aircraft high prior to spin entry. One day I pushed the rudder pedal to initiate a spin and the ground failed to appear in my windshield. More by instinct than reason, I applied opposite rudder to stop rotation. We seemed to be suspended, looking at the sky. I lifted my head backwards and caught sight of the ground. We were upside down. After what seemed an eternity, the nose dropped and the checkerboard of farmed fields rushed at me with unusual speed. My airspeed indicator showed 119 knots – a speed to be avoided in this aircraft. I remember saying "Is this it?" before I pulled out of the dive, the G-force driving my brain into my seat in the process. Obviously, I started spin entry with the nose too high and paid the price for it. I remember Gloria, my then flight instructor, telling me later "You don't have to stand her on her tail!"

Clearly, flying is not for everyone. You must want to be up there. Being a visitor in the air is a privilege. Flying opens vistas denied to those who do not venture in the air. Once you start flying you are always looking at the sky. You read the moods of the air, the clouds and the mountains and must accept the fact that sometimes they are not at home to visitors. Being scared of flying is not a disgrace. We all have different levels of tolerance and should stay within our comfort zone. Fortunately, life is so generous with the variety of interests it has to offer, we can all walk away satisfied.

Much is said, with exaggeration, about the dangers of flying. In truth, flying is safer than driving a car. Of course, a pilot has added responsibilities to himself and others and must keep up his knowledge and flying skills. The preponderance of time spent flying is immensely pleasant.

Flight training concentrates on safety and prepares you to deal with emergencies, should they occur. It is like learning the language of a distant country you may visit one day. If you practise what you were taught and keep learning as time goes on, you will not be caught wanting. If you consider yourself above it all, "the ground will come up and smite you'" as the saying goes. You are the master of your own safety.

It goes without saying that I have made my share of mistakes and was lucky to get away with them. Flying an airplane is rather like being in the company of a woman. The airplane, like a woman, does not want to share your attention with other things. In my case, she forgave me twice, but only just.

On one occasion I was about to land when I chose to take a photograph of the runway ahead. My train of thought was interrupted and I bounced the landing because I failed to cut power to idle. After two more bounces I corrected my mistake and, feeling badly about the poor landing, applied full throttle to go around for another circuit. By this time we had eaten up almost half the runway. When we reached flying speed the aircraft just would not rise in the air. I checked my instruments, eating up more runway. Still no response. As there was little runway left and there were houses beyond it, I considered doing an intentional ground-loop, that is, give full rudder to cause the airplane to cartwheel and crash on the ground. One more try, I decided, before choosing a fiery end. She rose from the runway sluggishly, like a very heavy person from a deep couch. We did achieve climb, albeit at a very slow rate. Still, we were flying! After clearing the buildings ahead I checked again and found that I had omitted to do what Geoff had hammered into me: advance the throttle and the carburetor heat knob with the same hand motion. Distracted by picture-taking, I forgot the latter, consequently I had less than full power. I was lucky.

Even less excusable was the take-off with full flaps when, distracted by a lot of chatter on the radio, I failed to complete my pre- take-off checklist. In practical terms I had the air-brakes on. While the engine sounded all right as I climbed out over the lake, my airspeed was pathetically low. At first I thought that my airspeed air intake (pitot tube) was blocked by a bug. When I could not achieve normal climb rate I realized that I had offended again by inattention.

Both these episodes served as a lesson. Both times I skirted disaster of my own making. The airplane taught me that it gives little margin to less than full attention. I was humbled. Perhaps the most telling lesson was that the confidence acquired by flying hours must still pay its dues to the discipline of flying. The price of freedom is discipline.

"There are old pilots and there are bold pilots, but there are no old bold pilots" goes the saying. It serves as a reminder that the principles of flying are absolutes, laws of nature which permit no short- cuts. The bold pilots I have known who showed off their skills close to the ground, boasted about flying in bad weather, bragged about their instrument-flying abilities, are all dead.

Gravity, the final judge, gives no rain-checks.

I fly for esthetic reasons. I revel in the sight of hills, valleys, fields and mountains. Looking down from high altitude you see land as God made it, for there is little evidence of man. While flying alone, I often sing aloud for the sights I see fill me with joy.

10,000 FEET ABOVE VERNON

In the Rogers Pass at 7500 feet

One autumn afternoon strips of cloud hugged the hillsides and cloud layers traversed the sky at various levels. The light was grey, the air still. The chatter on the radio told me everyone was flying at low altitudes. I became curious to find out what the real ceiling was that day. I climbed. It was a remarkable experience, rather like venturing far out to sea. As I climbed higher, the valley with its lakes, hills and fields turned dark, as if seen through smoked glass. I continued climbing. The darkness below turned darker still. The air was calm. Suddenly I found myself in a snowstorm at the bottom of a dark cloud. I relished the moment. I felt I was in the presence of a gentle, kind being, vastly greater than myself. It was almost as if I had been allowed a glimpse of eternity. A good feeling pervaded me. Yet I knew that the privilege was mine only briefly, so I put the nose down and returned to the airport. As I made my way home, and even now, years later, I wonder what it was that I saw and felt that day, but that experience alone made all my flying worth while.

I have to admit to a fascination with clouds. I revel in their shapes, their colour, am awed by their silent dignity and intrigued by their unknown, distant destinations. I enjoy nothing more than the company of billowing, white clouds in the early morning after a night of heavy rain. Flying in a little airplane among those gleaming towers is an experience like no other. It is not a place to stay, your visit is short, so every moment precious.

I also encountered some detractors who spoke disparagingly of my taking up flying late in life. They said, not kindly, that I was always out to prove myself. From my perspective life is short and I want to taste the whole of it. You must make your own agenda in life and not be held back by the crowd. I gain no satisfaction from staying with the group. I want to strike out in new directions, experience the excitement of reaching for the unknown, meet my own joys and failures face to face. If that is what they mean by "proving myself", that is the way I like it.

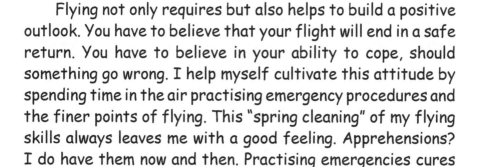

Flying not only requires but also helps to build a positive outlook. You have to believe that your flight will end in a safe return. You have to believe in your ability to cope, should something go wrong. I help myself cultivate this attitude by spending time in the air practising emergency procedures and the finer points of flying. This "spring cleaning" of my flying skills always leaves me with a good feeling. Apprehensions? I do have them now and then. Practising emergencies cures them every time.

The interesting aspect of flying emergencies is that sometimes they come unannounced. One moment you are full of the joys of flying, the next you feel that jab under the ribs that tells you things are not well.

One June morning I was flying over mountains when the air was still and the sun high, the kind of morning when I'm as likely as not to be singing aloud. As I made a routine check of my instruments I noticed the alternator warning light glowing a bright red. It was like finding a snake in a drawer.

The prescribed procedure is to turn the alternator switch off, then turn it on again. If the red light stays on, one must head for the nearest airport, as the battery is not receiving charge. The engine purred sweetly and I felt no threat. Within a minute or so I smelled smoke. I then thought I might be heading for trouble, so I called my home airport on the radio. There was no response. The smell of smoke became stronger, so I turned towards a valley some five miles away where I could make an emergency landing.

A pilot in an emergency can put out a "Mayday" call but that applies to serious malfunctions. I did not think a "Mayday" call was appropriate under the circumstances, so I opted for a lesser declaration:

"Pan-pan, Pan-pan, Pan-pan......Cessna 172 Golf Mike Oscar Victor (giving altitude and position) Have smoke in cockpit. Intend emergency landing in (giving position) area."

There was no response. I was a good twenty minutes away from the nearest airport. With the engine running fine I concentrated on maintaining altitude as I had to clear a ridge ahead before I could lower the nose for an emergency landing. The smoke in the cockpit grew a little worse.

Suddenly I heard a loud bang up front, like an explosion. The engine was still running fine and my dials showed normal readings. So this is what it feels like to be shot down by the Red Baron, I thought.

I put out another "Pan-pan" call. This time it was answered by Rob, a flying instructor at the airport. I told him I was preparing for an emergency landing. The smoke was clearing from the cockpit.

"Is your engine running alright?", Rob asked.

"Running fine."

"You have an electrical problem" said Rob. "Shut down all your electricals and head back to the airport. We'll keep Runway 23 clear for you".

"You mean shut down the radio, too?" – I did not relish the thought of being out of touch for the next 15 or 20 minutes.

"ALL your electricals", answered Rob.

I did as Rob advised. To be able to speak to him was a great comfort. With the radio shut down, there were no such comforts left. Still, I had altitude, a set objective and the engine was running faithfully. What was more, I was happy to find that I was functioning well. I still had no idea what had caused the smoke and the loud bang but knew that if I maintained altitude I would have time to select a forced landing site, should things go awry. My only cause for concern was that the airport was on the far side of the city. I could not make a gradual descent, since in the event my engine failed over the city I would have no landing option. I just had to maintain altitude and make a steeper landing approach than I was accustomed to making.

My reading about flying came to my aid. I once read that in a situation such as I was facing one could lose altitude by making S-turns in descent until the right position for landing was achieved, all the while keeping the runway in sight.

Flying without the radio was a lonely experience. Still, the landscape below was familiar. When I reached the city I was too high for a straight-in approach. The S-turns were a great help. After shedding altitude I had the runway in the accustomed perspective for landing. My left knee began to shake as I crossed the airport fence but I made a good centre-line landing.

The problem? My alternator seized. Meanwhile, the propeller shaft continued turning, causing the belt to burn. Eventually the belt snapped, hitting the engine cowling with a bang. The Red Baron was not flying that day.

My only other emergency arose when, inadvertently, I flew into smoke one day. Slash-burning in Idaho brought a lot of smoke into our region. The change in visibility was so gradual that I failed to notice it until I had no visual reference to the horizon or the ground. I might as well have been in a cloud. It was a nasty surprise. I remembered the "178-second" advice about life expectancy while flying in such conditions and set up the emergency procedure I had been taught.

I noted my heading, then waited for the second-hand of the clock to reach the next half minute. At that point I began a two-minute left turn, holding the aircraft level. The idea is that at the end of one minute the aircraft will have made a 180-degree turn, heading back to where it came from, out of trouble.

That was not to be. I had been flying with the sun behind me and now that I had turned around all I could see was the milky whiteness of the smoke, lit up by the sun. I was not better off than before.

There was only one thing to do. I turned away from the sun and began a descent at 300 feet a minute, keeping the wings level by reference to the artificial horizon on my dial. As long as the wings were level, I would soon break out of the smoke. I did. I sang all the way home.

Flying has been generous to me. It gave me a dream fulfilled. It taught me how precious the prize if only I have the resolve to pursue it. It gave me a measure of myself in a way few other endeavours can. To me, flying has also become a mode of self-expression, a way of seeking and finding excitement, beauty, wonderment and peaceful reflection. Encompassing space within the limits of my little airplane I am aware of my own insignificance, my transient visit to a world so vast, my mortality. I now look at passing clouds with the knowledge that I have been up there, in their midst and for those brief moments they welcomed me, let me fly with them, gave me something to remember.

In the context of my life the time I spend flying is fleeting. It is natural to want to record it, make it last. For this reason I always fly with my camera. I look for beauty that can be seen only from the air: silhouettes of mountain ranges viewed through a curtain of mists, reflections of clouds in water, the moods of the seasons in the valleys and the hills, the forbidding profiles of stark rock faces, the turquoise lakes concealed among haughty peaks and my fellow fliers, the clouds.

One day, while practising emergency landings, my instructor Gloria made a chance remark I never forgot. "Don't be afraid to put the nose down," she said. Gloria was an excellent pilot and I enjoyed teasing her, saying her mother must have been a fighter pilot. Flying alone, I started playing fighter pilot. Over a highway I would put the nose down, sighting a truck in the middle of my windshield and keeping it there while diving on it, shouting "rat-tat-tat". I would then pull out of the dive while the hapless trucker continued on his way, not knowing he was dead.

My early photographs from the air were spoiled by the intrusion of aircraft parts, the engine cowling, the wing strut, the compass. Then I thought of what Gloria had said and my problem was solved: I had to take my pictures through the top of the windshield where the view was unobstructed. To do this, I would select my target and dive on it steeply, keeping it centred during the dive, just as I did playing with the trucks. This enabled me to get two clear shots with a loss of not more than 600 feet. Not only effective, this mode of picture-taking is exhilarating in the extreme.

On a mountaineering trip to Nepal my daughter Sari took a flight to the Lukla airstrip, one of the highest in the world. She told the pilot that her father flew a Cessna 172. The pilot replied: "It does not matter what airplane he flies, if he flies he's a pilot!" This chance remark stayed with me and still gives me a sense of belonging.

My graduation cap and gown came when I took my youngest daughter Zsuzsi on a flight over high mountains. On landing, she gave me a hug and said: "You're a good pilot, Dad!

Chance remarks, both. They have probably forgotten them by now, but I never shall.

One day I ran into Hedy, a little old lady friend, who always asks direct questions.

"I hear you are flying an airplane" – it was more a question than a statement.

"Yes, I am."

"How many propellers?", she asked, her forehead wrinkled in a frown.

"One."

"That is too dangerous", she said, shaking her head. "Hedy, the heart has one propeller!"

SILENT KILLER

Prejudice is a silent killer which overtakes you without warning. You recognize it only by its deadly effect. Generations have suffered, millions have died in its path. A global pandemic, it is wholly amenable to cure. Alas, its persistence through the ages is sign of a fatal flaw in mankind.

In old Europe the social layer in which you were born was the layer in which you stayed for life. The concept of "nobility" imposed a barrier to upward movement which, for all practical purposes, could not be crossed. The "stigma" was permanent. Non-acceptance of one who tried would remain lasting. This may yet be the case in societies, few as they are, where aristocracy in some form is still a recognized and accepted entity.

In countries which have discarded the monarchy and its aristocratic trappings, there emerged a new form of aristocracy: that of wealth. Even here, hard distinction is drawn between *old money*, represented by families which have held wealth for many decades and, in their own eyes at least, are the deserving "monetary aristocrats", and *new money*, to which belong the "nouveau riche" who have the means but not the "pedigree of lasting success" in managing it in a reputable manner. Thus, winners of jackpots or merchants of doubtful credentials remain pariahs until and unless fading memories finally erase their upstart image. The Kennedy family name was linked with questionable business practices before achieving distinction. Some even regard the family as American "aristocracy", but for an unfortunate clause in the Constitution. The Bronfman family name was dragged into reports of illegal liquor trading and tax evasion during Prohibition. They went on to establish a reputable business empire and are respected patrons of the arts.

The way a man speaks is audible evidence of his origins, his level of education and his place on the social scale, all three being determinants of his prospects for success. Prospective employers are leery of appointing one who is not "one of us". This may be a way of protecting group cohesion and image, of defending against an unknown or little known quantity, fear of introducing a new element, or, as they see it, a new threat, into the mix. They also foresee that a man disadvantaged by his accent or dialect, even if well qualified for a role, would seem, to them, out of place and therefore undesirable in the social context usually binding those who work together. Thus, the barrier is not breached.

A foreign accent can, and often does, invite unfavourable reaction and resentment. I remember being called a "bloody foreigner" by people in England who clearly felt threatened by the arrival of strangers who would then become their competitors for jobs and food. Even in Canada, a bastion of tolerance, or so we like to think, my oral examiner in pediatrics, upon hearing me speak, turned to his fellow-examiner and in an aggravated tone remarked "What is the matter with this examination? Are there no Canadians taking it?!" By then I had served in the Canadian Army and was a citizen of Canada but not in his eyes. This slanted view of the "hyphenated Canadians" is more prevalent than many would admit.

If the above pitfalls of prejudice entrap those who share more or less similar physical features, differences in appearance mediated by racial differences will often magnify resentment. We just need to recall the legacy of hatred in segregated societies. In wartime Europe the Nazis portrayed the Jews painting exaggerated Semitic features in their hate propaganda posters. "Chink", for Chinese, is a derogatory term remembered by most Britishers, while the German propaganda of the "threatened Yellow Peril" did its best to depict the anticipated enemy with slanted eyes. The well-established Canadian designation "WASP" (White Anglo-Saxon Protestant) is monument to a deeply rooted view, in the eyes of some at any rate, that those who belong are the cream at the top. All this may be changing now but do not expect such notions to be cast away just because a black man has been President of the United States. Derogatory terms hide under a thin veneer, fly back and forth, and will be with us in the long term. We have heard them all: "Hun", "limey", 'kraut', "rooinek", "frog", "wog", "nigger", "yid" and "goy", to name but a few.

Like swirling gusts of wind, prejudice may move in any direction, seek any target. A world-renowned Harvard scientist I knew elected, towards the end of his working life, to devote a year of his time to teaching at a university in the southern United States where the students, and most of the faculty, were black. His very gesture was testament to his lack of bias. Yet, when he sat down in the cafeteria for lunch no one would sit with him.

Religions can divide mankind, alienate groups, and so serve the cause of prejudice. The Mormons call the rest of us "gentiles", while Islamic fundamentalists describe non-Muslims as "infidels". "There is but one God, Allah and Mohammed is his Prophet" is a basic pillar of their faith, invalidating all other concepts of deity. But then intolerance of the Roman Catholic Church of all other "religions", with prospects of Hell for the unbaptized, is yet another flagrant example of bias. Under Communist rule those of non-working class origin were designated as "class aliens" or even as "class enemies", to be deprived of privileges accorded to fellow citizens. Among Hindus I met who observed the caste system I detected silent disdain towards members of lower castes.

Civilized societies espouse majority rule. People want to be governed by those who are acceptable to most of the electorate. Even under those circumstances, opposing opinions elicit acrid dissent and uncivil condemnation. Election campaign TV commercials here in Canada and in the US are replete with examples of attempts to discredit legitimate opposing candidates. In our "civilized" parliament much of the public's time and money is wasted on building bias, digging up dirt, manufacturing resentment against each other by parties across the political divide.

Some may disagree but it is largely true that people tend to associate with those of a similar body type. Perhaps this is because those of comparable build tend to pursue kindred interests and create progeny in their own image. Those of lean and athletic physique seldom relate well to grossly overweight individuals. The former may even judge the latter silently for "not taking care of themselves". Sitting in a narrow airplane seat beside an overly corpulent fellow passenger on a long flight will not fuel warm thoughts of friendship. Those of great height will shun the short.

Physical handicap, even in this enlightened age, may quietly exclude a person from an invitation list. Examples of prejudice? If not, what?

Resentment of the better fortune of others is a powerful motive for ill will, vilification and, sometimes, violent action. Sadly, much of envy is generated by those who have done little or nothing to earn the rewards and comforts for which they so resent others. Rather than being spurred on to higher effort to achieve ends for themselves, they sit on the sidelines jeering.

Life is an arena of challenges and we are free to seek our own goals. Of course, to achieve success we must be prepared to give up something in return: our time, our comfort, our leisure and, in some cases, our youth, in the process. Life would indeed be dull and devoid of incentive if every participant in a race received the gold medal. Whatever our circumstances at the start, we do have a chance of improving our lives if only we have the desire. Improving our own lives gives us a chance to improve the lives of others.

A group has an unspoken wish for self-preservation. This is not unlike the way the human body guards against the invasion of foreign proteins: an allergic reaction serves as a warning that the visiting foreign protein (allergen) is not compatible with the resident protein structure. If, despite the warning of an allergen the same protein is again introduced, a strong reaction, rejection, anaphylaxis, occurs. This mixing of proteins is the crux of the problem in tissue transplantation.

The metaphor is crudely applicable to the genesis of intolerance by one group of another. The more the minority, the "foreign protein", asserts its apartness, stays "in your face", the more it is perceived as intent to dominate, therefore the more violent the rejection. This is why groups which integrate quietly are often accepted while those that flaunt their apartness evoke such negative response. "When in Rome do as the Romans do". There is wisdom in that. Champions of "multiculturalism" tend to ignore its inherent pitfalls. It works only if the newcomers arrive with the intention of becoming part of the society which welcomes them. If they come with stipulations to suit their own ends, enduring resentment and conflict will follow. When picked up by a lifeboat, be thankful, make no demands.

On the human scale, we have to rise above the raw model of the behaviour of proteins, more so in a world that ease of travel has made smaller. Alas, we are dealing with slow evolution in acceptance and adaptation, which knee-jerk accusations of "racism" and "intolerance" will do nothing to accelerate.

If only we accept the other man's right to be there in our midst, if only we can rejoice in seeing his life improve, if only we learn to live with the notion that his progress is our progress also, the collective progress of mankind, there is a chance that one day we shall walk side by side, in harmony with our fellow man.

As long as we isolate ourselves, withdraw into tribal enclaves, place ourselves above others, exploit each other, plot mutual revenge instead of discussion to allay bygone wrongs and fears, we shall continue to douse the flame of hope for a world where prejudice is a memory of the past.

In Bali, Indonesia I met a fascinating young man: a foreign correspondent for a Russian news agency. His magnificent profile brought classical Russian writers to mind. We compared notes. I told him I had had a bad time at the hands of Russians in the Second World War. His reply:

"But I wasn't even alive then!"

Lesson learned.

At my own level, all I can do is to say to myself again and again:

THAT PERSON I DON'T LIKE:

MAYBE I JUST DON'T KNOW HIM..................

TO AUSCHWITZ - AND BACK

I avoid organized tours. They make me feel part of a herd. The information I receive may be limited, unless I am pushy enough to propel myself to the front, which is against my principles. Fitting in with the schedule of a group offends my sense of privacy. Besides, I really do not enjoy unwanted conversation with strangers, some of whom will stick like glue when they find a willing listener. Distractions like that I can do without. Good reasons to do my exploring on my own.

I grew up in Hungary during the Second World War, with its many horrors which I was lucky to survive. I never heard of Auschwitz until the war had ended. The reason was clear: the media were tightly controlled by the Germans during the war and we had access only to their version of events. Everything was reported in a manner favourable to their cause. Just an example: the German Army did not retreat all the way from Stalingrad; it merely "lost contact with the enemy". Not surprising, therefore, that when first reports of death camps reached leaders of the Jewish Council in Budapest, they refused to believe them.

As it happened, I knew a good many Jewish friends and neighbours who disappeared, never to return. Years later, a half-century after I had left Hungary, a personal pilgrimage to Auschwitz to honour their memory was a natural decision.

Yet I knew that on such a visit I had to acknowledge all who died there. To single out any one group would offend the memory of all others whose lives were also extinguished. It would make their sacrifice meaningless, as if it was of no consequence.

According to the map the distance from Budapest to Auschwitz was about the same as from the Okanagan Valley to Vancouver. At 4.15 one afternoon in May I left Budapest. My rented car was registered in Austria but I could drive it across borders.

At the border into Slovakia the guard looked at my papers. "Valami nem stimmel", (something does not add up), he remarked in Hungarian and disappeared into the booth. About an hour later he emerged, handed back my papers and raised the barrier without a word. Perhaps finding dollar bills in my passport would have expedited my transit. I do not subscribe to bribes. The drive that followed took me into the scenic foothills of the Tatra mountains, villages and small towns rolled by, their esthetic appeal marred by the plethora of red Communist Party banners bearing slogans. I had visions of picturesque hunting lodges or country inns with good food but along the entire Slovak stretch I saw no accommodation, nor even a gas station. My full tank would see me through to Krakow, my objective for the night. As we approached the Tatras, the scenery became beautiful: castles perched improbably on rock pinnacles while deep forests lined the way.

At dusk I arrived at the Polish border. The guard was friendly enough but after consulting his friends he demanded payment equivalent to 12 U.S. dollars, even though I already possessed a visa. Polish is a language which leaves me befuddled, so I just paid up in order to make use of the daylight remaining. I was now at the original border of old Hungary, before the Trianon Peace Treaty of 1919 ripped the country apart. Trianon certainly took the best out of Hungary!

Light failed rapidly in the mountains. The road was narrow, twisting and dark. All I could see was the path lit up by my headlights. Although I must have been getting close to Krakow by now, there were no lights in the sky ahead to show it was there. A makeshift sign down the road said "Krakow, 10 Km". Good news! I was tired and extremely hungry. As I reached the outskirts a modern sign stated "Hotel Forum". The place was lit up, contrasting with the gloomy surroundings. I pressed on, wanting to stay somewhere in Krakow itself. It was 11.30 p.m.

The "downtown" part of the city was poorly lit, lifeless. There was no traffic in the streets and no sign of humans.

After driving around in circles, finding nothing, I knew I had to return to the "Forum". A few more minutes, a shower and food!

At the desk I was informed that the hotel was full. There was no available accommodation and the restaurant was already closed. Alternative accommodation? Sorry, there was nothing. The clerk showed no sign of even trying to help. I was determined not to leave it at that, so I asked whether everyone who had made a booking had checked in. It turned out that two guests had not shown up.

"In that case, I shall take one of those rooms"

"But that is impossible!"

"Not impossible at all! It is now midnight. I am here in person.

There is no reason I cannot have the room".

The clerk continued to protest, so I asked him to call his supervisor. I explained my predicament in the most polite terms and he let me check in.

As for food, I was out of luck. The restaurant was closed. However, there was a disco downstairs where I might be able to pick up some snacks.

I was hungry to the point of distress. I found the disco, where loud music played inside. Outside the door a half-dozen young men, wearing leather bomber jackets, were hanging around, trying to steal a quick look inside every time the door opened. They were obviously boyfriends of the girls who were dancing within with foreigners and affluent Poles. Anxiety was written all over boys' faces. It was not my milieu. I disliked the atmosphere and besides, I never go into bars. There and then I decided that it was fitting to go to bed with my hunger, after all, the inmates at Auschwitz did it all the time. Opposite the elevator door on each floor sat a couple of weathered women, ostensibly to make sure that none of the disco girls would continue the dance in the guest rooms.

Morning revealed a fine view of Krakow from my window, with the Vistula river close by and the cathedral beyond. After an early breakfast I set off for the only gas station for miles around and patiently awaited my turn in the long line-up. The city looked drab but style and old glory were all too evident.

At last I was on my way to my destination along rolling terrain with rich farmland in every direction. A sign indicated Oswiecim (Auschwitz) straight ahead. It was hard to believe that an evil place like that could be concealed in such beautiful countryside: a deadly snake in a well-tended garden.

Following the road signs I finally arrived at Auschwitz Camp. I had to sign up for the tour, luckily the group was small. We walked under the famous "Arbeit Macht Frei" sign, to find a row of red brick buildings. These were used to house prisoners and were the site of "experiments" by SS "doctors", such as immersing people in ice cold water to see how long they would last. Along the corridors there were hundreds of photographs of Polish prisoners, heads shaved, wearing striped prison garb. In the lane outside there was a rail set on three posts where 18 prisoners were simultaneously hanged, side by side, after a failed escape attempt. The rest of the captives had to stand and watch, as a deterrent.

The other buildings housed glassed-in displays of mountains of human hair, mountains of spectacles, toothbrushes, prosthetic appliances of every kind, even suitcases with the owners' names painted in large, white letters.

The last of the red brick buildings was the death block. The courtyard was bordered by tall brick walls, the "wall of death" at one end, where prisoners were shot. In the middle there stood the gallows. The building housed cells, some for the purpose of torture. I tried one: it was so narrow that you could not sit and so low you could not stand. Even a couple of minutes so confined were most uncomfortable. I stood in the courtyard, away from the others for a minute, to link up in spirit with those for whom the brick walls and cobblestones were the last things they saw.

Once outside again, we visited the gas chamber which was surprisingly small, with openings in the ceiling where the "Zyklon B" cyanide was introduced. Close by there was a small crematorium with only a few ovens. Right opposite were located the living and recreational quarters of the SS staff. A somewhat satisfying exhibit close by was the gallows on which Hoess, the Commander of the camp was executed by hanging.

The tour took less than two hours and concluded with a grainy movie showing conditions in the camp before its liberation by the Soviet army in January 1945. Then the show was over.

I left the place with the feeling that, despite seeing mountains of hair and toothbrushes, I had not really seen Auschwitz at all. There had to be much more. This camp was far too small to exterminate humanity on the large scale I had expected. The limitations of a guided tour were all too plain. Such displays as were shown dwelt only with the suffering of Polish prisoners. As always, I had to do some exploring on my own.

Poland is not an easy place to find your way around. Signposting is not generous. People speak Polish and not much else. English, at least at the time I was there, was hardly spoken. I thought of trying to enquire in German, but having seen "the wall of death" and mountains of hair on display I thought it unwise. Then in a flash I recalled the bits of Russian I had taught myself in 1945 and drove up to a man working with a shovel down the street and addressed him:

"ГД Е Birkenau?" (where is Birkenau)

"Dva kilometer", replied the man, tossing his head over his shoulder to indicate the direction. All it took was four words and I was on my way.

The rail spur leading under the SS guard house into Birkenau extermination camp had my stomach in knots. It felt as if I myself was reaching the end of my road. I parked the car and walked right between the rails, remembering all those who passed through there. I recited the words from Dante's Inferno: "Abandon all hope, ye who enter……". I listened for the barking of the dogs, for harsh words of command, for cries of distress and was deeply moved.

There was no sound. A light breeze was blowing and the grasses bent to its will. In this enormous cemetery of mankind's innocence I was alone. I did not even see birds flying. Just silence, silence so heavy that it was hard to bear. This was a pilgrimage, not a sightseeing tour. I walked along the rail spur to where it forked into two. This is where countless people, men and women, young and old, even little children, had arrived at their final destination. I felt their uncertainty, their fear, their pain of separation from those they loved, from life itself. I blessed my good fortune that I could be alone in this place of anguish. I thought I should speak, say something appropriate, utter a word of solidarity in this soul-destroying silence but my throat felt tight and I could not articulate the pain that was inside me. This was a time to suffer, a time for silence, a time for just...feeling. Everything I might say would be trite, nothing I could utter would be fitting epitaph for what had happened here. Something in all of us died here, the shadow of this place will follow us as long as we live. This is where we learned that nightmares do become realities.

I stood on the collapsed roof of the gas chamber which the Germans blew up in their retreat. I visited the long huts still remaining, most of them had been burned down. The door was open, swinging, creaking to the dictates of the wind. Inside the huts there was a long bench in the middle with holes cut along its length on both sides for lavatory use. At each end there was a fireplace. To both sides three tiers of platforms marked the space where the prisoners slept. No privacy, no comforts here. A sign, leaving no doubt, declared:

"Ein Laus, dein Tod!"

(if we find a louse on you, you're dead)

Along the barbed wire perimeter at intervals there were elevated platforms for the armed guards to discourage those who would try to flee. The main guard house was empty, with free access to every room. In the centre section above the rails a recording was playing to an unseen audience. It played sounds, commands in German, yet more evidence that the Germans kept tidy records.

This was the most meaningful pilgrimage of my life. The fact that I was alone, with no other human in sight, no one in earshot, no one to distract my thoughts was a gift. I left Birkenau humble, grateful……

…………and hungry.

I elected to drive back to the Czechoslovak border without returning to Krakow. I drove through villages without encountering restaurants or even cafes. Starving the night before had been ordeal enough. I needed food.

The town of Wadowice (birthplace of John Paul II) came into view. I drove around, desperate for food. Finally, in what appeared to be the centre of town, I found a restaurant. I knew language would be a problem. The place was full as I entered. Not a table empty. I addressed a waiter hurrying back to the kitchen. He shrugged his shoulder, he could not understand English. I knew that trying to read the menu would bring no joy, so I followed him into the kitchen. I thought I could point at the dish of my choice and just wait for a table to become vacant. The kitchen was tidy and the aroma of food inviting. There was a row of trays from which the orders were served on to plates. Beyond it, several girls were busy with their work. I looked at the trays critically and noticed that one of the girls, on her lunch break, was sitting at a small table with an appealing dish on her plate.

This was a time for sign language, so, making sure that both the waiter and the kitchen staff had my attention, I pointed at the dish in front of the girl, then pointed at myself, showing that was what I wanted.

A roar of laughter arose from among the pots and pans. They all thought I wanted the girl. I joined in the laughter and by appropriate gestures I made it clear that, although the girl was pretty, it was the food I wanted. The waiter nodded his understanding.

When I returned to the restaurant a young man approached and guided me to a small table on which table cloth and cutlery had already been set for me. He indicated that he and his friends did not need the two tables at which they were sitting and were happy to accommodate me. This was a most kind, generous gesture on their part. I thanked them sincerely. Unfortunately they spoke only Polish, so conversation was not possible. The waiter had already arrived with the item of my choice (not the girl) and within a minute one of the boys came in from outside with a bottle of beer they collectively bought for me. My gratitude knew no limits. I was humbled by their kindness. When I left I persuaded them to accept a substantial U.S. dollar bill from me, so they could enjoy drinks together. I shall never forget their kindness and generosity. Gestures like theirs are a gift from life.

The journey back to Budapest was memorable. Beautiful scenery escorted me along the way. A few hundred yards from the Polish-Czech border, around a bend, in a "No stopping" and "No photography" zone I saw a stunning stand of tall evergreens and decided to risk incarceration to record them on film. With the camera around my neck I got out of the car and, pretending to be emptying my bladder, I faced in the direction of the trees and took photographs. For good measure I emptied my bladder also.

My fuel gauge told me I needed gas. There were no gas stations along the way, not in Slovakia, nor in northern

Hungary. I used every measure I knew to save fuel. By that time I was hungry again. The experience of the night before was repeating itself. In Budapest everything was already closed and again I starved, as had the hapless victims of the concentration camp I just left behind.

Unlike them, I did return home safely.

FAREWELL JANOS, MY BROTHER

The sons of two sisters, Janos and I were born just two months apart. Of all the people in my old family, he is the one I have known the longest, and from the dawn of my recognition he was my brother. Of the family, apart from our children, only Janos and I were left, two aging boys in their eighties, dwelling on, living on, relishing the memories of days long gone.

Janos died at dawn yesterday. My loss is immeasurable, pervading, incapacitating. I feel as though I have become blind, for it was through Janos that I saw and understood myself. From an early age I took my cues from him because he was wiser, more agile and more daring than my childhood self. Hesitantly I followed him climbing trees that I had thought were beyond my reach. He taught me to ride a bicycle. He was able to ride round the block when I was still crashing and scraping my knees. Together we rolled the fallen leaves of autumn into pungent cigarettes and coughed our way through the experience. We attended the same school. The only fight I ever had, and it was against my nature to fight, was with a boy who said or did something to Janos I did not like. I came out winning and felt smug about it.

We spent most of our free time together until the war came. After every bombing raid my first thought was for phoning him to see if he was all right. In the air raid of April 3, 1944 I came close to losing him. In the aftermath I visited him and we stood at the rim of a large bomb crater across the street where nothing was left of a house and a friendly family who lived there only hours before. Death left its calling card and we did not talk much for the rest of the day.

When the land battle arrived and the Russians closed the ring around the city Janos and his family moved to the presumed safety of a villa in the hills of Buda, across the Danube. For two months our lives were just an exercise in survival from one day to the next, dodging bullets and shells, getting inured to the fear of danger stalking the streets. The maelstrom of the conflict invaded all our senses, sometimes even making us forget the hunger and the cold which were our constant companions. Whether we survived or not was not up to us. Faceless men with guns ruled the day and the night.

We lost contact for several months. I recall how, when the shooting stopped and the guns fell silent, my first thought was of Janos, how I wanted to know that he, like us, had weathered the storm. The river was a frozen mass of ice, the bridges had been destroyed by the Germans, who, on the orders of Hitler, were told to defend the city "to the last man". They held out for two whole months in the two high points, the Gellert Hill with its ancient Citadel and the Castle Hill, a mile to the north which, for a while, gave them a field of fire until the battle closed in, the big guns became useless and the fighting was reduced to small arms, even hand to hand.

On our side of the river the war continued not with gunfire but rape and plunder. On the Buda side, the 40,000 strong German garrison, running out of ammunition and any hope of survival, broke out, attempting to penetrate the Russian ring, flooding, screaming through the narrow, centuries old streets. The Russians waited for them and there followed a massacre through which only a few hundred managed to get through. These were quickly rounded up. The house in which Janos stayed was commandeered by the Russians. The prisoners were marched in one by one at gunpoint by Yuri, the cook, they were interrogated, then led out and shot.

The bodies were then laid on the road shoulder to shoulder, a truck then came and flattened the heads into pancakes. Janos witnessed all this. By this time his shoes were worn, so he took the boots off a headless body, his feet finally dry, free of snow and ice.

It took a couple of months before a makeshift bridge was built, making it possible to cross the river. We had a joyful reunion, one of the happiest days of my life.

Later Janos came to visit me on our side of the river. We walked and talked, exchanging experiences. A Russian soldier emerged from among the ruins, aimed his rifle at Janos and made him give up his boots. Janos was left with the Russian's broken, leaking boots. The law was the whim of the man with the gun.

Eventually we returned to school, or a semblance of it. After the turmoil of battle it was difficult to adjust to being students again. Not only that, but the new Communist government, helped by the armed Russian presence, decreed that the house in which Janos and his family lived was much too big for them, so an arbitrary wall was built in the centre of the house and a Communist family moved in. Summary arrests and executions made everyone live in fear. When the "free" elections" gave the Communists only 17% of the vote, people who were unlikely to vote Communist were simply taken off the voters' list.

At this time our school, run by Benedictine monks, came under government scrutiny. It was clear that such a school did not fit in with government plans, so a good reason had to be found to discredit and, eventually shut down the school.

Two boys were accused of killing a Russian soldier and were duly hanged.

It was at this time that I was fortunate enough to leave for England on a year's scholarship. As it turned out, I left Hungary for good. The dangers for the boys left behind continued to grow. Our immediate circle of friends, Janos included, was accused of an "armed conspiracy aimed at overthrowing the government". They were arrested and severely beaten, in an attempt to extort false confessions.

While the trial was being prepared, Janos was taken out of the cell one night and driven to Military Police headquarters where he was met by an engaging, sympathetic officer who told him it was a downright shame to waste his young life in prison. He told Janos he was welcome to order food, any food, from a nearby restaurant, have cigarettes and coffee in plenty and stay till morning. There was only one stipulation: he gave Janos paper and pencil and asked him to write down everything he knew about our Head Master, a Benedictine monk. With this, he left.

In the morning he was back, fresh and smiling, and asked Janos whether he had done his homework. Janos said he had, and handed the man his sheets of paper. The pages were blank. With this, the smile had left the man's face and he said to Janos "You know you have just signed your own death sentence".

The trial followed. Three more boys were hanged, the rest, including Janos, were given sentences of 10-15 years. Janos spent the next five years in a crowded prison cell under harsh conditions and frequent beatings. The cell was occupied by men of thought. Our Head Master taught his cell mates English, a pathologist lectured them on the natural history of diseases. (one of the boys visited me at The London Hospital after their release in 1956. As we sat in the pathology museum for a quiet chat, he pointed at a jar behind me, saying "look, that's infarction of the spleen!").

Janos's health suffered greatly as a result of beatings, privations and brutal prison conditions. Beatings of the soles of his feet left him with a permanent limp. His lungs were compromised, his exercise tolerance minimal. Clearly, his body was broken. But not his spirit. Where others might emerge with thoughts of resentment, vengeance and hate, Janos cultivated love for his fellow man. Despite having served time in prison, forever after he was ineligible for jobs other than the most menial kind: hospital orderly or mortuary attendant. His life, his prospects were taken from him for good.

He did not leave it at that. Deeply religious, he took a degree in theology and worked as an assistant to the parish priest, helping to preach and conduct weddings and funerals. The congregation liked him so much that they asked specifically for Janos to conduct such ceremonies. This was much to the chagrin of the parish priest, but that is how it was. Janos also felt he had to give spiritual comfort to incarcerated prisoners and became a recognized prison spiritual advisor.

Broken in his body but defeated in spirit he was not. His conversation was sparkling, informed and it reflected his deep faith. His sense of humour was a quiet undercurrent to his views on the world, most of all on himself. He was not a victim, though all his life had been taken from him by force. Yet the greatest force dwelt within himself, a kindly fountain of understanding, of unconditional love.

No beatings, no guns, no death can conquer that.

Sleep well, Janos…………good night…………

CHARITY BEGINS - AND ENDS

I am wary of charitable causes. I am kind-hearted and ready to help those less fortunate than I am. I always gave generously to those who had an unkind hand dealt by fate. It made me feel good to know I had made their lives easier.

Or had I? Experience has told me time and again that far from easing their suffering it was only my own conscience I relieved.

Corruption is a pandemic with no apparent cure. The corrupt line their pockets with money that would feed the poor. The children of Africa starve while their rulers live in palaces. Food delivered to areas stricken by drought is sold by those in power, or left to rot in warehouses if the poor cannot pay. Medicines sent to some mission hospitals are not dispensed free, as intended by the donors, but often sold to patients even by "Christian" missionaries. Moneys collected worldwide for earthquake, tsunami or flood relief are trapped in cobwebs of bureaucracy. Those responsible for their distribution are not challenged to account for their disposition.

Here in Canada charities are persistent in their fundraising: expect to be approached by the "disease of the month" appeal, the Spring, Summer and Annual fund drives, not to mention Christmas and Easter drives. Scrutiny of some charities reveals that only a small portion of the funds collected goes to the cause they represent. The percentage spent on administration and fancy offices is enough to discourage the most willing donor.

Then there are the endless raffles where you are persuaded to buy a ticket in a good cause. Embarrassing if you don't. Now I turn them away. Why? I used to participate in an annual racing event, always concluded with a banquet attended by some 800 people. Everyone bought a ticket at the door for a draw. The prize: choice racing equipment. Two years running the winner was part of the inner circle of the organizers. I was then invited to a cocktail party with 18 guests at the home of one of the principals. Two people in the room were past winners of the draw. Do the math!

In my travels I came across poor people in Indonesia who, I thought, gave me a chance to dispense charity where it was needed, without the interposition of a middle man. My satisfaction was immense. True, in trying to help I always steered the recipients toward a goal, which if they achieved with my help, would raise them out of poverty. The money I gave away was for education with the goal of exchanging living in a shack for comfortable quarters in a house.

Disappointment soon followed. The education money was spent on cosmetic purposes: skin whitening, braces for the teeth and laser eye surgery to eliminate the need for wearing glasses. None of it was spent on education. Money intended for accommodation was lost at a gaming table. Eva gave a Christian man a generous gift to enable him and his family to visit his parents on the island of Sulawesi at Christmas. The money was spent on a second motor bike instead. I could go on.

One time my generosity was rewarded. A girl I knew in Bali had a big thyroid tumour. I advised her to see a surgeon. She said the surgeon wanted $1,000 U.S. for the operation. With ten days of my intended stay still to go, I decided to return home ten days early, gave her the money and came home. Three days after my arrival home I suffered a heart attack. Had I stayed in Bali I would not have lived to see the ambulance arrive. Was it reward from a higher power? I shall never know, but ever since that time I have been less dismissive of things I cannot fully explain.

Of course I continue to give to charitable causes but I choose them with care. Remembering my days as an intern in the East End of London, I always give to the Salvation Army. In the crypt of a bombed church nearby there lived an assortment of derelict, homeless men who regularly got drunk on methylated spirits. When the bottles were empty they would smash them and proceed to carve each other up. They arrived in the Emergency Department bleeding profusely from facial wounds, wrapped in the stench of their unsanitary existence, not to mention the repulsive smell of their methylated spirits hangover. They were in no fit state to discharge. Hospitals were full of sick patients and would not accept them for even temporary care. The Salvation Army always agreed to take care of them.

I also have great respect for the organization *Médecins sans Frontières*, doctors and nurses who put themselves in harm's way in trouble spots of the world.

Organized charity has become an industry and it secures a comfortable living for many who work in the field. We need to satisfy our collective conscience by giving, yet exercise good judgment lest our satisfaction turn to regret.

ORGANIC SOLUTIONS

Barely awake at 7 in the morning, we were treated to a lively earthquake. Like severe turbulence in a high-flying jet, it began without warning and tossed everything about for a good twenty seconds. Then, as quickly as it started, it stopped. While we had experienced earthquakes before in other places, even a 5.9 here in Bali, this one was postgraduate stuff, not the hesitant rumble that makes the Venetian blinds quiver, but a deliberate, persistent shaking, involving the ground, the walls, the roof and the very bed in which we were lying.

A few minutes later it returned. This time it lasted a minute or more, the roof, the walls, the floor moving visibly, the "I huff and I puff and I'll blow your house down" kind of commotion. It was a marvel to witness from the comfort of one's bed. In truth, we were so taken by surprise that, instead of making a quick exit to the outdoors, we just watched the spectacle unfold. It is a wonder the roof stayed on but the woven roofs of houses in Bali are just placed on top of the walls, so there is a lot of "give". In the back of our minds we were thinking "tsunami" but it never came. Now it is true that we have our own protocol in the face of a threatened tsunami: there is a hill at the end of the beach about three hundred yards away. We place our clothes and footwear in exactly the same place at night, so we can find them in a hurry. If we walk fast towards the hill and begin the climb once there, we can reach palm tree height which we consider a safe elevation. It takes us eleven minutes. We just hope we get fifteen minutes' warning.

Now, as for tsunami preparedness in Bali, it leaves a lot to be desired. The locals simply avoid the topic. If you try to call attention to the possible consequences, they do not want to hear. The gods, they "know", will look after them. In truth, there are three sirens on the island, in the tourist-populated areas, each with a limited audible range of about a kilometer. The protection they offer is limited. In the low-lying areas, heavily populated by local Balinese, such as Jimbaran and the busy Ngurah Rai Bypass which skirts it, there are no sirens. Jimbaran is also where we stay. The truth is that the southern part of the island is barely above sea level, the roads are congested with traffic they were never designed for and, warning or not, not many would be lucky enough to reach high ground.

I actually got involved, wrote to Made Mangku Pastika, the Governor of the island and others in decision-making positions but received no response. Finally a well-meaning friend wrote back to say that no immediate intervention is indicated. "Bali will solve the problem *organically*", meaning: things will work out somehow, all in due course. Interestingly, later the same indifference visited me while on the island and I gave up thinking about the matter.

We learned that the earthquake we had experienced was 6.4 on the Richter scale, with its epicentre in the sea, not far to the south.

While in our world disasters or even impending catastrophes precipitate media frenzy, networks swarming the area with reporters on the ground and helicopters hovering above, in laid-back Bali none of this pertains. The *Jakarta Post*, one of Indonesia's leading newspapers, reported on the steadily increasing, serious encroachment by the sea on the beach at *Lebih*, a small Balinese village on the east coast of Bali, where each year for the past few some seven meters of sand have been ceded to the sea on the beach. The Bali papers made no mention of this, nor was there evidence of any follow-up action, nor even seeming concern, regarding this fact.

We made a trip to Lebih to see for ourselves. We got there about an hour after high tide. As we walked the passage between rows of *warungs*, really just flimsy vending stalls, toward the sea, the street was still wet under our feet from the water now receding. The vendors sat by their wares unconcerned. On the beach obvious erosion was evident, the sea undermining the sandbank on which the outrigger canoes were resting. A little *warung* at the water's edge still open, its foundation itself undermined, clientele still sitting around in dispassionate conversation, just spending the time of day. No sense of urgency here. We bought drinks for everyone and settled in for a chat. Yes, the sea level has been rising but really, no problem. Just a little more difficult to haul the boats up on the shelf but the sea is the sea, what can be done about that! The gods will look after everything. Smiles were the order of the day and they quickly followed up on jokes.

Nearby, seated on a *bale bengong*, a platform with a thatch roof which serves as a venue for ceremonies and events of all kinds, the *orang suci*, a holy man, was conducting a ceremony. Clearly, this must be for expiation of the gods of the sea, we reasoned. Incense burning, a little bell ringing out now and then, a circle of villagers seated around the edge of the platform with expressionless faces, another "organic solution" for a calamity in the making. I did not wish to intrude, so I took a telephoto shot of the gathering. It was only later, when I reviewed the picture taken that in the ring of worshippers I saw a man texting on his electronic gadget, with a couple of women looking over his shoulder. Perhaps he was connected to a higher power.

There is no question about the sea levels' rising. Several years ago the Japanese government paid for reinforcement of the tourist beaches in Kuta and Tuban, on the west side of the island as well as in Sanur in the east. The project lasted a couple of years. Sea walls were erected with limestone ferried by barges from uninhabited islands in the archipelago and countless tons of sand taken from beaches that could spare it. Little breakwater islands of rock were deposited offshore in vulnerable areas. The initial results looked good. On later visits there were signs some of the new sand had already washed away. A local man told me that at high tide the water often washes over the new sea wall.

In our own world we stage conferences and invite *savants* to theorize about events we cannot fathom. We talk about carbon footprints and impose taxes to flagellate ourselves in attempting to avoid the inevitable. Having been to the moon and back and now having landed on Mars, we think we can control Nature itself. The Balinese just sit back and let things take their course. They do not rush to solutions of doubtful promise. It will turn out somehow, they know, and problems will solve themselves, "*organically*".

As said Robert Frost:

*"We dance in circles...and suppose.
The secret sits inside......and knows".*

A footnote to all this, in 2012 we again visited Lebih beach. The little *warung* where we had spent time with the locals is no more, as are a couple of old men we chatted and laughed with only last year. The government had imported a large quantity of rocks and built a sea wall, a couple of miles long. Little woven trays of flowers and rice are placed daily on the recently laid boulders, offerings to the gods. As for the new wall, the locals are not impressed: "give it seven years and it will be gone", they say. Even while we were there, at high tide the sea washed arrogantly over the new sea wall.

We visited the sea wall built in Kuta a few years ago. The walking path on the top of the wall is regularly awash by wave action at high tide. As we walked along, we found a spot where the sea had already found a plane of cleavage between the rocks and begun to prize them apart.

Lebih beach erosion, 2011

The beach at Lebih, east Bali, has suffered progressive erosion in the past decade. Structures built on the sand have been undermined and destroyed by rising waters. Beached boats and dwellings are retreating inland. It is a losing battle for the low-lying land against the relentless sea.

LETTER FROM HONG KONG

LETTER FROM HONG KONG

The erstwhile British colony of Hong Kong is all China now, dating back to the handover in 1997. The waters of Hong Kong harbour lie between Hong Kong Island and mainland Kowloon. It is an astonishing layout, with tidal waves of humanity, colour and noise. The drivers in Nathan Road, a major thoroughfare in Kowloon, display unfailing patience in a mass of traffic. The traffic itself is well-channelled and gridlock is a thing of the past. Here and there, in patches of green, people go through their Tai Chi routine, oblivious of the multitudes. Whereas back home we are used to seeing maybe 10-15 people waiting at crosswalks, in Hong Kong there are 50 or 60. A Milky Way of neon lights competes for attention. Shops are open at all hours, there is industry all around, all in a sea of smiles, patience and good temper. At night the city sparkles like a diamond.

Then there is Hong Kong harbour with a hundred water craft, ferries, junks, cruise ships, cranes and motor boats, all criss-crossing each other's path in a manner defying all logic and, to be sure, all maritime rules, showing that when masses of humanity live at close quarters safety is more a matter of good judgement than the fruit of complex regulations. The harbour, a wonder to behold, resembles a child's drawing of a busy waterway. The fare on the Star Ferry is one of the best bargains in the world.

There is a concert hall near the Kowloon Star Ferry terminal into which I happened a few years ago, only to find it packed. Everyone was listening, as if in a trance, to a group of musicians from China, playing Chinese instruments, producing a sound of haunting quality. It was an act of bonding in sound, bonding with music springing from their roots, echoes which had dwelt only in their subconscious, affirming their identity.

A strange place, HongKong, still very British, orderly and well- mannered, a comforting legacy of old colonial rule. Yet it is unmistakably, irreversibly Chinese; history, like water, finding its inevitable course.

Despite the influence of Beijing the people of Hong Kong, predominantly of Cantonese origin, are not about to turn their backs on their Cantonese heritage. In a fine shop I was served by a tall, elegant assistant, clad in a well-tailored suit, wearing a stiff white collar and a Windsor- knotted tie. I asked him: "now that you are a part of China, what language do you speak, Cantonese or Mandarin?" He pulled himself up even taller and in a confident voice affirmed: "we are Cantonese, we speak Cantonese!". In a multi-level shopping mall twenty young people were waiting for the elevator. They did not crowd the door but stood in single file, awaiting their turn.

Flying into Hong Kong is a new experience, more comforting than in years past when the big 747 would turn on to final approach to Kai Tak airport with the wingtip heart-stoppingly close to the balconies of high-rise apartment buildings where the washing was set out to dry on long wooden poles. The airplane had to drop onto the runway quickly to enable it to come to a stop because the far end of the runway was marked by the sea. The new airport at Chek Lap Kok is one of the wonders of the world, thanks to British engineering know-how. In a master stroke the British, before they left, made sure of getting the contract for construction, an engineering blueprint to be studied in times to come.

The sheer size of the new airport is intimidating as you approach it. However, once inside the doors directions are clear, obvious and logical. A particular gift is the check-in counter for seniors: no line-up, no waiting, you check in and are on your way. Sparkling shops light your way in the concourses, a spotless subway train will whisk you to your departure gate and everywhere you meet an abundance of space and light.

Across the harbour from Kowloon Hong Kong Island is presided over by Victoria Peak. High-rise buildings of bold design declare confidence in the future. The striking buildings of yesteryear have been upstaged by new construction on land reclaimed from the sea. In open spaces solitary Buddhists meditate in the lotus position, while nearby a group of zealous young people belt out Christian songs at an impromptu gathering. Powerful motor ferries line up for their turn to pick up passengers, the boats heaving in the swell created by the constant traffic. Their stop is all too brief, they are hardly away from the jetty before the next boat is already boarding its human cargo. The pace is compelling, it never slows, time at a premium does not allow one to stop and smell the roses.

A detour to the container port is a revelation in itself. While in other places one may see lively activity, the scale in Hong Kong is staggering. Mountains of containers await transfer to ships through a veritable forest of cranes, working at all hours to send the work output of unseen millions across the oceans to the four corners of the world.

If good humour is a language, it is alive and well in Hong Kong.

Gaps in the mutual vocabulary between vendor and customer are easily made up with smiles, nods and simple hand gestures even where English is not shared. The vendors are eager to sell and good-naturedly expect polite bargaining. "I ask eighty, you say fifty? How about sixty?...... Okay!" and the merchandise is handed over with a smile. In their view both vendor and buyer must be happy when a deal is concluded. The abundance of merchandise is astonishing, the temptation to buy hard to resist.

The typhoon shelter at Aberdeen has changed radically in the last twenty years. At one time the water was home to hundreds of boats tied up side by side, fishermen and their families living out a tradition as water-dwellers. Children of fishermen became fishermen themselves, ordained to do so by age-old custom. The women created a home on board while the men were out to sea. Many of the children never went to school, knowledge of the sea was all they needed for survival. So it continued in the long term.

When China took over the law insisted that all children receive a school education. Children left the boats and went on land to attend school. The more time they spent on land the less inclined they were to return to life on the boats. Soon the parents moved also. The typhoon shelter is still there but homes on the water are few now. Towering high-rises on shore attest to the shift in population from the water to land.

There is considerable division of wealth in Hong Kong. Only four percent of the city's inhabitants live in houses they own. Thirty-three percent own condominiums and the rest pay rent to the four percent of house-dwellers. Still, by all appearances the Hong Kong market must give the world's luxury car makers a reason to be satisfied.

A visit to the Hong Kong Stock Exchange, the Hang Seng, is an anticlimax of sorts. One would anticipate seeing a trading floor with feverish activity, shouts and frantic gestures over a floor strewn with paper, the pulse of business driving a vast economy. There is none of that. The traders sit motionless in concentric circles behind a glass wall, with no apparent talk, all gazing into their computers: a frozen orchestra playing silent music without a conductor.

Always enjoyable is eating in a Hong Kong restaurant. It is obvious that the Chinese love eating in groups. The choice of food is endless. Their conversation is spirited, their meal spiced with laughter and is a good example of how you can have a good time without alcohol. People who relish food so much are not likely to make war.

On the way to the airport one thought occupies me: when will I return?

LETTER FROM BALI

LETTER FROM BALI

My stay in Bali is coming to an end. I look back on a time which made me forget the date, let alone the day of the week, filling my mind with thoughts of sunlight and shadows, colours, sounds of the surf, laughter and the touch of a gentle breeze brushing my skin. These give me comfort and make the prospect of a long winter at home bearable. Be that as it may, by cyclical experience I have grown accustomed to a life where the sun always rises and sets at the same time, where there may be breezes but hardly a wind, where the air is so warm that your attire is nothing beyond what modesty dictates, where Nature delights your eye with an unending variety of tropical flowers while the song of birds visits your ears.

Not that my stay in Bali was plain sailing. As it happened, I contracted some kind of 'flu virus while on the island. Being sick in a hot climate is a punishing experience and I felt its full weight. So stricken was I that I reluctantly went to the clinic to see a doctor. In the waiting room I felt wretched. And then I was ushered into the presence of the doctor, a gorgeous, nubile, 26-ish brown-skinned apparition, the very sight of whom cured me instantly. Well, at least for a moment. She turned out to be a really good physician, she asked the right questions and did all the right things. Finally, she gave me a prescription and asked me to wait for her written report.

And then the axe fell We are all familiar with the situation where the patient is crushed by bad news received from a doctor. This was my turn. The report began: "*generally looks well......*" This was the final blow. I expected her to write "*....exceptionally well-preserved septuagenarian of athletic build and a glint in the eye under silver hair, bravely fighting a nasty virus......*". After all, in Bali I walk on a carpet of compliments about my youthfulness, my zest, when compared to Balinese men of much younger vintage. "*...generally looks well*" was more than I could bear. Vanity is the last bit of flotsam to cling to in the shipwreck of old age. I was sick for two more weeks.

The island of Bali is undergoing slow but unmistakable change. After the bombs of 2002 many prospective visitors stayed away. The apparent trend to recovery that followed was halted by repeated bombing in 2005, much to the detriment of the tourist-dependent economy and the wellbeing of the population. In the wake of the explosions the tourists fled to the safety of their countries and those of us who stayed had the island to ourselves. It was a sad sight to visit restaurants, once lively with chatter, music, waiters scurrying about with delicious food, now empty, yet the tables all laid, with candles burning, waiting for diners who would not come. When a young waiter was commended for the well-laid tables and smiles he wanly replied: "……… look behind our smiles and you will find the tears".

Everyone knows that the bombings were perpetrated by Muslim fundamentalists. Yet, although their livelihood was virtually wiped out, the Hindus acted in the most civilized fashion. There was no rancour, revenge, retaliation, no "eye for an eye". They simply turned the page and went on with their lives. Surely a blueprint for supposedly civilized nations which engage in endless games of pingpong killing in the name of tribal or national pride, to say nothing of religion.

Another change is in the making. While up to a few years ago 95% of the population was Hindu, now Muslims are increasingly more visible. Daily you see women wearing the Muslim *jilbab*, which actually gives a charming air to many. I saw only one woman clad in a *burqua*. Alas, I noticed her too late for a telephoto shot and she refused my respectful approach for a photograph.

The reasons for the Muslim presence appear varied. Clearly, Indonesia, being the most populated Muslim country in the world, would want to bring Bali into the Muslim fold. Certainly the more radical elements of the faith regard that as a mission in itself. Then there is the envy of Bali from all around the archipelago because Bali has been so attractive to the rest of the world and has benefited greatly by it in the past. Couple this with the lack of employment in neighbouring east Java and the impoverished islands of Flores and Timor and the reasons for a shift in population become clear.

Now it is obvious that the attraction of Bali lies in its Hindu character. The innate charm of Hindu people, their gentleness, their mode of dress and their unending, colourful ceremonies stamp the island with an ambience which attracts foreign visitors and makes them want to return. However, in the very Hindu fabric of Bali lies also its weakness: to the practical onlooker there appear to be just too many Hindu ceremonies. Just to name a few, they have ceremonies lasting one or more days for births, first haircuts, the equivalent of christening as we know it, the filing of teeth which is a form of "coming of age", the full moon, marriages, invoking the gods to protect the motorcycles, the main means of transport on the island, tributes to the gods of the sea, deaths and cremations, to say nothing of the visit of the gods to the island every 210 days which puts a hold on productive work for the best part of two weeks. It must be mentioned that the offerings prepared for such ceremonies are costly in terms of income, more so if neighbours and friends try to outdo each other. No wonder Governor Pastika of Bali proposed simplifying ceremonies to keep expenses in line with incomes.

Now, if you imagine yourself having to schedule work for your employees with all these interruptions it will come as no surprise that Muslims move in to fill the gap. They are available and willing to work even during their holy month of Ramadhan when they fast, with no food or drink from before sunrise until after sunset. The Muslims are tidy, whereas, sadly, the Hindus often allow today's accumulated holy offerings at the door to become tomorrow's litter.

There is another aspect of Hindu thinking in Bali which, with the greatest respect I have to call excessive credulity . Take the case of the sinking of the ferry boat between Bali and the island of Lombok. The boat sailed late one evening with its cargo of cars, trucks, crates and many passengers. It never arrived. The local Hindu population, including senior hotel staff of managerial rank, believed that the gods caused unusually severe wave action to send the boat to the bottom of the sea. At home we call it overloading.

The son of a Hindu villager we know, now about 10, suffers from asthma. It is a common ailment affecting children and mercifully many outgrow it before they reach their teens. So they were advised by the physician who also treats the occasional acute episode. The family is more inclined to believe the religious healer who resides on the slopes of Mount Agung, at 11,000 feet the tallest of Bali's 15 volcanoes. The Hindu population flocks to this shaman who is popular, yet not accountable to anyone for his opinions. Our friends consulted him about the boy's asthma. He thought deeply about the problem and finally asked if they had a rabbit at home. When the answer was "yes" he told them to kill the rabbit and feed it to the boy. When the treatment failed he suggested they killed their dog and make the boy eat its flesh. The asthma continued. One evening the father found a stray bat flapping around the room. He caught it and served it up fried. The boy's asthma continues.

Late one night the boy's father felt feverish, could not sleep. His wife, worried, not least because she had to rise at four in the morning before a one-hour motorcycle ride to work, telephoned the shaman who, to his credit, had his cell phone turned on at 1 a.m. The shaman listened to the story of the husband's insomnia and had no immediate answer to the problem. He promised to call her back. "There is a ghost in your house", he stated when he called back a few minutes later. ".....but don't worry, I spoke to the ghost and told him to leave". Thus reassured, the family slept peacefully thereafter.

Then there is the invasion by developers, mostly from Australia, who are buying up land and building luxury villas that sell for improbably high prices to those who can afford it or are foolish enough to be absentee owners to buy. Foreigners cannot own property outright in Bali.

Buying expensive real estate in a country where the political future is hardly assured, where random laws may dispossess foreign owners of their acquisitions, on an island located in the "ring of fire" of volcanoes, a shifting sea bed and the threat of tsunami seems hardly a wise investment. There is evidence in the press of resentment by the population toward these developments which, little by little, gnaw away at Balinese-owned land and visibly change the character of the island. Shades of Maui, Hawaii, where the trend went too far and is beyond reversal.

A friend told of the valley where he lived not so long ago. Rice terraces hugged the slopes where the local farmers worked tirelessly by day. In the heat of noon they would retire to little huts to shelter from the sun overhead. Somewhere in the valley the soothing sound of *angklung bamboo,* a percussion instrument made of hollow bamboo trunks, floated on the air, as if affirming that this was the land that would house them, provide for them while they lived and receive them after they died. Now the iron pegs of foreign surveyors dot the land and the *anklung bamboo* is silent. You hear many stories like this in private but effective public opposition is lacking. After all, money rules and despite government attempts at eradicating corruption, decisions are made at a level beyond the reach of the islanders who are too preoccupied with constant dialogue with their Hindu gods.

Corruption is a disease alive and well in Indonesia. Government commissions attempting to wipe it out come and go. The latest of these attempts was derailed when the chairman of the commission, reportedly involved in a love triangle, was arrested for murder. Corruption creeps into every aspect of life: the doubtful destination of the five billion dollars' worth of aid received after the tsunami of 2004, the issue of permits for expensive land development, the recent misappropriation of visa payments by immigration officials at Bali's international airport, down to the policeman who stops a vehicle for a "check" on the understanding that the driver may proceed after handing over a banknote. It is a way of life, tacitly accepted. In a minor collision, onlookers will urge the drivers to leave the scene before the police arrives because the policeman will offer a choice between filing a "verbal report" on the scene for, say, 50,000 Rupiah ($5 US) or a formal report back at the police station for twice that amount. As a taxi driver once put it: "our police are merely symbolic". Back home in our "holier than thou" world corruption is swept under the rug by rank or the services of a good lawyer.

I had an interesting encounter with a man working in a consular job for a European country in the Philippines who told me: "in the developed world you pay a bribe to buy something illegal; in the Third World you do it to expedite something perfectly legal". Food for thought.

In recent years a new tourist has appeared on the scene. Hailing from north Asia, he is noted for his free-spending ways which spawned the emergence of signs in a new alphabet, inviting those who can read them to pleasures the island has in store. A restaurant-owner tells me these new arrivals frown upon ordering a single shot of vodka, no, they buy a whole bottle, often another. They celebrate with abandon and tip with an easy hand which leaves the rest of us in the dust. Once under the influence, they can become gregarious or boorish. An obnoxious, well-oiled member of this fraternity refused to pay his substantial bar bill and when the hapless waiter asked him to reconsider, he demanded to see the manager. The duty manager happened to be a charming, gentle, slightly built, Balinese man. The intoxicated guest looked at him with contempt and then asked to see the REAL manager! The security man was summoned, a well-built, fine-looking young man, Bali's erstwhile martial arts champion. His exceptional power and commanding presence were masked by a most engaging, polite manner. "How can I help you, Sir?", he addressed the aggressive guest. The guest, instantly aware that he was outgunned, rose, saying "Um...I'm just leaving......", with banknotes cascading from his wallet even as he spoke.

Not long ago I attended a cockfight in the Karangasem village of Timbrah, north-east Bali. Strictly speaking, cockfights are banned now but they are a tradition deeply rooted in the Hindu culture of the island. As part of religious observations, cockfights survive. All over Bali the crowing of roosters is part of the background of sound. These roosters are well-looked after, strong, large birds, whose only purpose in life is to do well in the ring and thus justify the bets placed on them by eager punters addicted to gambling, a serious problem on the island. A fine young man I once knew failed to turn up for work one morning. It turned out he had hanged himself in the night because he had amassed gambling debts he was unable to pay. The suicide rate in Bali is disturbingly high.

I must dwell on this young man, Wiliasa, for a moment. He had an engaging manner, a strikingly good face, he was slow of movement, ever ready to smile. He spoke deliberately, everything he said seemed to be taken off a shelf of well-considered thoughts. Meeting him was always an occasion. My last conversation with him was of circles. The circle is the most perfect of shapes, he said. "Just look at how Nature loves the circle: look at the Sun, our source of life; the Moon, the fullness of which we Hindus celebrate every passing month as it brings light to our nights; the roundness of the earth on which we live and die; the arms of a mother held in embrace......" He would let me ponder these images and not let his words intrude into my thoughts before, like water into dry ground, they had time to be absorbed. My departure from the island was drawing close and as he turned to leave I called him back to give him, as I always do, money in appreciation for all he did for me during my stay. In the most polite manner, almost apologetically and certainly most unlike others on the island, he declined to accept it, as if a monetary gift would taint our joint recall of conversations we had enjoyed. His death by his own hand during the night that followed disturbed me no end, for I could only guess at the torments that visited him in his last hours. I felt I would have helped him without hesitation if only he had bared his plight to me. His ethics allowed no room for a bailout by others from a predicament of his own making.

A stranger in a village can find the hall where the cockfight is staged by just following the crescendo of human voices and the cry of the birds waiting in cages to enter the fray. *"Morituri te salutant!"*, the salute of gladiators in Roman times, is here the crowing of the cocks, visibly excited even before they enter the arena. The birds are taken from the cage, matched by the roughest measures of weight and size before they are placed on the floor. The betting is already under way, large sums risked on the outcome. The spectators' voices rise from "forte" to "fortissimo" as the cocks, still held by their handlers, are brought face to face, feathers ruffled, while the four-inch knives are taped to their legs. A few more face- to-face thrusts bring the animals to a state of anger where they can be let loose on each other to ensure entertainment for the onlookers. The birds circle each other tentatively, then leap into the air, knives thrusting forward. Every thrust brings forth a thunder of cheer from the crowd. Sometimes a single thrust leaves a bird dead on the floor. Sometimes a long series of exchanges leaves one wounded, limping, unable to fight back. He is taken from the ring and is killed on the spot. Now and then powerful animals are unable to kill or maim each other and give up the fight due to sheer exhaustion. The handlers then put them into the same cage where their proximity still engenders aggression but lack of space prevents them from fighting. Remember, a lot of money is at stake and the gamblers must be satisfied. They are then taken from the cage and made to fight again, to the death.

Attending the cockfight was not something I enjoyed. I did not witness style or beauty. I felt sorry for the animals. An enduring impression was the unbridled fury of the crowd, urging on the contestants, not unlike that of spectators in a hockey game back home, enjoying a fight between players. Strangely, at a bullfight in Madrid, Spain, many years ago, I had the same pity for the bull but could not deny that the drama in the bullring had its own ritual, its style, its stark, dark beauty. We should not condemn violence in other cultures while finding excuses for it in our own.

Why do I keep returning to Bali? The answer, in short, is that in Bali I found the perfect end-destination, a place where the pace is of my own choosing, where the climate is kind, the people friendly, the food delicious and the nights peaceful. When I think of the millions living in the midst of wars, hunger and thirst, struggling to survive from one day to the next, I consider myself blessed, safe from all that. It is the people of Bali who make the island what it is, a paradise of sorts where I do as I please, where I can set my mind wandering and, even if none of it finds its way to paper, nothing is lost, there was no destination in sight and the journey was reward enough.

VILLAGE ON THE MOUNTAIN

The temperature is a soaring 32°C this humid Bali morning. After only a few minutes your clothes, wet with the moisture your body sheds to keep cool, cling to you uncomfortably while you go about your business, soaked in your own sweat.

Up ahead a car is waiting, waiting to spirit us away to the highlands where the climate is temperate, the air comfortable and cool. Inside the car the harsh reality of the air-conditioner greets us with its sudden, almost rude relief, like a well-meaning but rough handshake. And then we're off, along the tree-lined Ngurah Rai bypass, named after the Balinese hero of the war of liberation from Dutch rule half a century ago. We pick up speed but are nevertheless passed by swarms of motorcycles flashing by us now on the right, now on the left, cutting in dangerously, weaving through traffic, squeezing through openings so narrow that the onlooker is left breathless. And so it goes, mile after mile till we are well out of town.

North of Sanur, flanked by the sea on the right, rice fields to the left, we settle down to cruising speed until the time comes to turn off into an improbably narrow but nevertheless paved road, winding itself through the ricefields where mosaics of land, rice in varying stages of cultivation meet the eye in every direction. Some fields are under water, with newly planted shoots in neat rows, in others growth is well under way; here and there the land, having given up its crop, lies fallow.

In the fields, tireless workers, wearing the picturesque hats of Asia, bend to their exhausting work. I ask our driver what their pleasures are in life, noting they are constantly at the job, away from amenities which back home give us comfort, a change of pace in our daily round. He smiles wanly and informs me that these people find comfort, pleasure, meaning for their lives in watching their crops grow. Harvesting the rice is the culmination of their effort before the cycle begins anew with planting the next crop.

Here and there along the road the rice, already threshed, is laid out on large sheets to dry, our respectful wheels barely skirting it. No industrial drying machines here, that is what the sun is for. Scarecrows, narrow sheets of cloth, shiny strips of tinfoil attempt to drive away predators that might damage the crop but everyone knows it is all futile: whatever the loss it is just the cost of doing business. We pass women carrying loads on their heads, bodies fully covered to escape the sun, men carrying sickles, *sabit*, the tool of the rice worker, their faces a document of a harsh life under the elements. Time to pause and muse about the ingenious gadgetry that back home eases our way through what we call hard work.

Imperceptibly, the road begins to climb, away from the sea. We are now heading for the high country where lush vegetation attests to daily visits of rain. Trees line the road, cover the hills above and populate the deep gorges cut by meandering rivers, the arteries of life for the ricefields below. Ever more tortuous, the road presents new vistas, new surprises at every turn.

The blue of the sky is now obscured by a canopy of green. As we ascend to gain a bird's eye view the land below transforms into a vast rice-bowl wherever the eye cares to look.

From the comfort of the car, village after village passes by. The thatched headgear of family temples rises above grey stone walls lining the road, with slate roofs over the stylish homes within. The gods are everywhere and the decorative arches, the *penjor*, symbols of the holy mountain, incline over our path. Dogs wander about aimlessly and chickens pick morsels from wayside heaps of scrap. Young boys dart out of narrow alleys on bicycles, old women walk the roadside with heavy loads on their heads, small groups of men sit on the steps of *warungs*, the village shops, in endless conversation.

We step out of the car into the cool air of the highlands. The master of all the hills, Mount Agung, the sacred volcano, is clothed in cloud, near enough but remote, not receiving visitors. A rough path, now climbing, now descending, crosses a fast-flowing stream, an important source of water for the village up ahead. A woman, knee-deep in the river, bends over a flat slab of rock, busy with her laundry while a stray dog looks on, cooling off in the stream.

The village itself is an island in a sea of rice, no more than a few houses here and there, connected by a footpath. One wonders how the inhabitants negotiate the rough, uneven passage on a moonless night. We come upon a house at the edge of the *sawa*, the ricefield where the crop is heavy with rice, waiting to be harvested. Inside the wall we find the basic family compound: a few small buildings, each housing a family within a family, sharing a courtyard with chickens running around and small children playing hide-and-seek.

Under a canopy, away from family traffic, handsome, big roosters in straw cages await their eventual turn in the ring at the cockfights. These are events attended by men, some of whom wager more money than they can afford on the outcome of each combat. Outlawed, except as a part of religious celebrations, the Hindu religion is never short of a religious excuse for ensuring that these contests survive. All over the island, thousands of roosters are kept in their handsome but confining straw cages, crowing the cries of a thousand dawns, as if knowing that their own dawns are numbered.

The houses are elaborately styled, with hand-carved and gold-painted wooden doors and window-shutters, every one a work of art. Within, the furnishings are sparse, elegant, comfortable. Even here, high on the mountain, you will find a television set, small children clustered around it, watching a program more rich in noise than in content.

The villagers are mostly friendly, although some view the visitor with reserve, at least until words are exchanged in their language. The children act coy before the camera until they relent and then ham it up with broad smiles. Thereafter an air of good humour prevails, the conversation a mutual enquiry into the lives of those who have come from afar and those whom they came to visit. Many nods of approval follow information which, in truth, will not be retained. Clearly, the essence of the meeting is not the knowledge we gained but that we met in a spirit of good will and mutual respect.

At the interface of different worlds we are all ambassadors for our kind, our kin, the colour of our skin, the language we speak and for the way of life we represent. Our humanity is our passport. Our brief contact with our fellow man from afar will endure only in the imprint we leave in our wake.

The farewells are fonder than the "hellos" which preceded them. Smiles, after all, are the universal language that recognizes no barriers, no walls. No treaties were signed, nothing was done or said that would change the world as a result of this brief encounter. Yet, all of us harbour a good feeling, a knowledge that we could never raise a hand in anger at each other.

Smiles and waves follow us back to the car, waiting on the road. On our descent we come upon a project where peanuts from Java are prepared and packaged for the local market. In large bowls they are washed in water from the river, downstream from where the woman did her laundry and the dog cooled himself. Inside, the peanuts are dried, roasted and bagged. All the workers are women, every heavy load carried on the head from one part of the plant to the other. The finished product looks and tastes good, one only needs to forget which way the river flows.

Embraced by the green landscape we make our descent back toward the sea. The sun is on its way down. Tireless women still carry their heavy loads with no destination in sight.

Men still toil on the land. Traffic intensifies as we get closer to home. Ahead lie the trappings of the tourist industry: the gaudy placards vying for attention, the invitations to expensive tours, the waiters outside restaurants beckoning to the passer-by to enter.

The peace, the silence of the distant ricefields, the clouded mountainside, the firm handshakes of the villagers, the innocent play of the children recede into memory. Inexorably, we are carried back to an environment ready to shed its way of life in exchange for hollow gains, ready to stamp out its character and reduce it to a copy of a thousand sunny, faceless seaside destinations in the world.

I cannot help feeling that the real Bali is up there on the mountain we have just left behind.

The Bali Times

Bringing You Bali, the Nation and the World every Friday

May 23, 08

THE PIRATES AND BEGGARS OF LAKE BATUR

The Pirates and Beggars of Lake Batur

By Nicholas Rety
For The Bali Times

LAKE BATUR, Kintamani ~ Every time I drive by the lush, green ricefields of Bali, its deep gorges and colorful villages toward the high country of Kintamani, I feel a new bonding, a new affirmation of my love for this island.

The lively mosaic of people at the market, the motorbikes zigzagging their way in the most improbable traffic and the constant reminders of the presence of the gods give Bali a charming, lovable, yet vulnerable character that I have grown to love.

More than in the Bali that is becoming westernized at a rate that sheds contact with its past, I find comfort in the Bali of old where a culture, respect for origins and roots still prevail. It is the remnants of the old traditions that set Bali apart from other dime-a-dozen sunny destinations around the world.

I once visited the Bali Aga village in Karangasem, a powerful trip back into the past, made memorable by the history, the customs and the simple way of life of its inhabitants. I left with a good feeling, carrying with me memories of their pride and mementos of their industry and artistic skills, which explained why I, native of a distant land, came to be in their midst.

Hardly surprising that when I heard of another remote enclave on the shores of Lake Batur, accessible only by water, reportedly living a life of simplicity, untainted by the crazy march of "progress," I at once decided to revisit the good feelings with which I left their counterpart in Karangasem.

The road toll on arrival at Kintamani would have bothered me less if there had been some accountability, in the form of a numbered ticket for its collection. The onslaught of aggressive vendors on alighting from my vehicle was something I had not prepared for but it did alert me for what was soon to follow.

On reaching Lake Batur I was asked for a cool Rp1 million (US$108) for the return 15- minute ride to the Bali Aga village. Being familiar with incomes and prices around the island, I knew this charge was exorbitant. Eventually I settled for Rp700,000, only because I had come so far and did not have time to come back again. The boat man knew this; he had met this situation before. He was neither fazed nor embarrassed.

Lake Batur was at its best, the ridge from Mount Abang to the south standing guard while the volcanic face of Mount Batur followed us across the water. There was no other watercraft to be seen, so we had the lake to ourselves, a special feeling in such a remarkable setting.

Soon the village came into view. Not the old Bali-style buildings but a treeless compound, devoid of greenery or flowers, placed like matchboxes elbow to elbow. There was total absence of color or of any attempt to beautify what looked like low-cost housing for laborers on the edge of some faceless town. A lamppost declared the presence of electricity in the village. A fleet of canopied motor vessels lay at anchor in front of a high wall at the water's edge. There was no dock to alight on and we got to shore by negotiating some precarious footing.

A small group of men met us on shore. No hearty greetings here! They had the body language of mosquitoes ready to bite. They ushered us uphill through rubble and garbage along a nonexistent path with tricky steps to a small temple with a handsome roof climbing skyward. The temple door was freshly painted but still unfinished. The self-appointed spokesman did not explain the absence of women from the scene, nor indeed the seemingly total absence of life around the village. At the temple door a bowl with banknotes of large denomination appealed to the visitor to make an at least matching donation. As I reached into my pocket the whole group closed in, eyes fixed on my wallet, some murmuring disapproval of what I considered a substantial gift.

The only female around, a little girl in the arms of her father, held out a pathetic hand for money, with a facial expression that would have qualified her as a model for any painting of The Last Judgement.

We were not allowed to walk around the temple. Instead we were led straight back to the wharf, if it can be called that, away from the village. It was as if the phalanx of local men were protecting the village folk from the scrutiny of outsiders who were welcome only to empty their wallets and leave. Repeated gestures here and there invited donations to fit men who appeared well dressed, well fed and not at all living a life of isolation, certainly not men whom time had passed by.

Re-embarkation was followed by a short trip to the burial ground. A few dugout canoes skillfully plied the waters, now restless with a rising wind. Upon stepping ashore, an elderly woman held out a hand for money. She must have tutored the little girl seen at the previous landing, her choreography and facial pathos matching.

The guide explained that when a villager dies, the corpse is kept in the house until such time, usually a week later, as the holy man decides that it is time to carry it to the burial ground by boat. It is then placed under a tent made of palm leaves and left for nature to take care of. The terrain does not allow for a cemetery. After a while, when decomposition and the mice have done their job, the body is lightly covered with earth.

Under the umbrella of a large banyan tree, we entered a dark space. The climb was strewn with garbage, cigarette packages, plastic bottles and wrappings, a child's sandals, all left where they fell. A small rise led to a row of skulls, arranged mechanically, face forward, an honor-guard on parade. The silence and the lapping of the water on the shore were the only signs of reverence for the dead. As before, a bowl well-filled with large- denomination banknotes awaited the visitor. Clearly, these bowls had been placed there just before our arrival because they showed no sign of the rain that had preceded us.

The palm-leaf tents came into view. There, within, one could discern a skull with the orbits already denuded of their contents. The teeth, survivors as they are, feigned a smile. The body was small, reportedly that of an old man. The toenails showed careful pedicure. I spoke a few words of respect to the corpse and gave a salute before turning to the bowl to make yet another donation.

Turning away from the palm-leaf tent, I found myself walking on a carpet of sundry garbage, mixed in with a variety of human bones, a femur, a female pelvis and other skeletal parts which appeared to have been discarded after severance of the skulls. I expressed my disapproval of such indignity to human remains and suggested to the 10 or so men that if each of us picked up 10 pieces of garbage right there and then, we could thereby honor the dead and turn the place into a more appropriate burial ground. I bent down to pick up a crumpled plastic shopping bag but was at once thwarted by a man standing next to me. This was sacred ground, he said, and could not be touched except by permission of the holy man.

Descending back to the boat, the inducements for tips took verbal form and hands were held out in supplication. I gave one man some money and told him to share it with the others. The old woman's hand was still out, as if frozen. I told her to take it back or else I would not give her money. One of the men, contrary to all appearances, complained about how poor they were. I quipped in reply that if I stayed long enough, I myself would become a poor man. By the boat landing there stood a brand new, Balinese-style building with a modern toilet awaiting future visitors.

The outboard motor came to life and we sailed off in choppy waters. Disappointed, I did not wave back. These were not timeless people living an old lifestyle. They did not practice any craft taught them by their elders. They did not wear their way of life proudly. If they all died tomorrow, there would be nothing left to show that they had made a difference. They were not guardians of an old culture which they nurtured for those who would come after them. They had lost their pride and become modern-day beggars. They turned death into an industry by the macabre display of corpses.

Not like their Karangasem counterparts at all.

I left the lake and its village of beggars with a bad feeling.

Later I learned that tour operators no longer visit this site, in an attempt to protect tourists from exploitation. As for the piratical boatmen of Lake Batur, two independent sources revealed that they have been known to stop the boat halfway through the crossing and demand additional payment before getting to shore.

I left the place without turning back for a last look. I tell myself the whole visit was just a bad dream.

When You Have Nothing, Yet Have It All

October 10, 2008

By Nicholas Rety
For The Bali Times

Were he not so inconspicuous I might not notice him. Not one to rivet my attention, he drifts through the landscape as a bird in flight, shadow of a passing cloud, a leaf driven by the breeze, a distant song dying on the wind. He appears again and again, oblivious of all as the world is of him, living in his own time, his own space. The view from my window, the landscape he inhabits, comes alive with his presence, a presence so transient, so trivial, so fleeting that I question why I notice him at all.

Then it dawns on me that it is his transience, his simple life in harmony with nature, his seeming dismissal of all the irrelevance the world showers on us, his seamless coexistence with the elements that mark him as a man of note. He embodies the reason why I come to Bali: not to recreate the glitter, the noise, the mindless waste of the world I call home, but to find refuge from all that. This figure in my landscape represents another way of life, a freedom from greed, a contentment with what is enough; maybe even a prescription for mankind's long-term survival.

He comes again, on time, as is his way. He heads for his hut among the palms, a flimsy refuge from vagaries of the weather, the heat, the rain, but his home all the same. He walks by his outdoor kitchen, no more than a corrugated iron roof on a few bamboo sticks, his stove a metal grate bridging two large stones, with pots and pans hanging above, dancing to the wind.

In the hut, his home under a thatch roof, a sheet of wood, raised above ground to stay out of the rain represents his bed, with a round log to serve as a pillow. An empty birdcage speaks his love of pet animals. Then, maybe it is a symbol that birds should be free, not caged. Against all reason, an old telephone sits on a wooden box beside the bed. Its bell is silent and the receiver has not heard a human voice for years. The ceiling is festooned with discarded plastic bags from expensive shops, now serving for storage of his few possessions. Fears of losing them? He has none. When you have nothing, you have nothing to lose. Any thief would disdainfully pass by this humble abode without giving it thought.

Outside, a washing line attests to his cleanliness. A new crop of laundry each day, faithfully washed and rinsed, awaits the sun. The grey tinge of shirts once white betrays the fact that he has no money for detergent.

The proudest, most striking feature of this dwelling is a flagpole, a tall shoot of bamboo, with a faded rag on top. It does not presume to be a national symbol. Yet it is a declaration that in its shadow there exists a human life, a soul clothed in poverty but not defeated by it, a pair of hands roughened by work but not disgraced by handouts, a man with no earthly possessions who is still richer, happier, more satisfied than the grumbling, greedy, anxious, insecure men of wealth in my world.

Oblivious to his penury, a contented cow grazes in the grass and calls him master. Broken shells of coconuts cut from the tall palms, fuel for the kitchen stove, are drying in the sun.

I never see him eat, never witness his private moments. He has no visitors, save for stray dogs which run back and forth, oblivious of his perimeter. There is no litter around him; he is as fastidious about his surroundings as he is disciplined about the routines of his day. Maybe in his mind he envisions himself occupant of a mansion or a villa giving on to the sea. Then I see him wander over to the cow and lead it gently to another spot, where fresh grass awaits.

Man and his animal living in peace, not dependent on outsiders. Man in a world of his own. Man in harmony with nature, the constant provider, adjusting to nature's moods and ways, taking in moderation only the essentials for day-to-day survival. He knows no excess; his wants encompass only the things he can carry. He is living a life that is, not a life that might be. His movements are slow and purposeful; he is not the prisoner of deadlines.

I approach him and start a conversation. He receives me calmly, with the dignity I was expecting. Ready to smile under the straw hat which never leaves his head, he gives me a firm, friendly handshake. His grip is that of a man who expresses himself by the toil of his hands. I am taken by his sense of humor, by his total lack of guile, his openness, the absence of any social obstacle between us, the readiness to be taken as he is, with his total existence on public view.

The one thing I do not see at a glance is what goes on in his head, his thoughts. He talks to me on the level. I feel at ease. Money he has none. His home is not visited by comforts. He is gracious, yet has no social status nor any prospects in view. His life is ruled by the rhythm of the seasons. Everything he needs he provides for himself. No wonder he is so positive, so self- assured. I have scaled a few peaks in my life but I am starting to look up to him. He knows, he understands something I do not. As a Western man, I feel I am still on the quest, searching for the answers to many unknowns, the enigma, the purpose, the futility of life; but here is a man who is content, serene, above it all. I am humbled by my discovery, glad that this inconspicuous, simple, almost invisible man wandered into my path.

And then I learn that this idyllic setting will soon cease to be his home. Someone has recognized the value of the land and he must go. One by one the palm trees will be cut down, the bamboo carried away. Plans do not allow for him to stay, for his animal to graze. The hut will be bulldozed away and the flagpole used for someone's building project. The flag, if it is one, will be hauled down and end up in a rubbish heap. For many of us this would be the end of the road: our home, the shelter to lay our heads when day's work is done, no longer a place but just an evanescent memory.

Scavengers are already at work, salvaging anything remotely useful at the site. The landscape is changing, the felled palms prostrate, lifeless. Where once an interplay of light and shade delighted my eyes, I now gaze into the vacant stare of an empty lot.

He shrugs his shoulder, shows no emotion. He even permits himself a smile. I am desperate to know how he can cope with all this, how he perceives his next step in life, the future or even just the next day. Too late. He mounts his bicycle and with a faint smile and a wave of his hand pedals down the road.

Filed under: Around the Island

May 29-June 4, 2009

BALI MORNING

May 22, 2009

By Nicholas Rety
For The Bali Times

JIMBARAN ~ Ferocious rain wakes me from my sleep. Surely the ocean has turned upside down, its contents dumped on the thatch roof above! A solid mass of water, the roll of a thousand drums, beats down on everything around me, the palms, the grass, the flowers. The pool in my garden boils white in the clash of water with water.

Maybe the sky is acting in anger and will not tolerate anyone, or anything, in the path of its mood. The thunder of crashing water drowns out all sound. The limits of my world are now defined by this wild river falling from the sky, for the moment putting everyone and everything out of reach. I am captive in a prison of water.

Then as quickly as it began, the rain stops. A few errant drops fall in its wake and, almost apologetically, late streams of water from the roof vanish into the ground. All is still once more.

I step out into the night. The raincloud is gone. Above, the sky is littered with stars, like diamonds carelessly strewn about the vault, their position constant, never changing. I am gazing at an infinity I cannot fathom, at an eternity I cannot comprehend. The surf, itself eternal, is the only sound I hear, its rhythm the only reminder of time. In all this endless mystery I am the one finite entity, unnoticed by the tireless surf, ignored by the haughty galaxies above. A magnificent solitude.

A hesitant rooster calls, receives no answer. The "town crier" of this lonely place, he assures me that all is well. The music of silence again envelopes the world, lulls it to sleep.

The darkness, the night promise to last forever. There is no moon, no breeze, no evidence of Man. No creature stirs. Time seems to stand still, marked only by the relentless beat of the surf washing the shore, a tempo so constant, so unending that to keep count is pointless. Time, after all, is relevant only to those who are passing figures on the scene, to whom eternity carries only wonder but no meaning. Dwarfed by the vastness of the ocean, the distance to a star, the infinity of Time, the limits of our understanding bid us to put aside our daily quest and seek solace in sleep, itself a mystery understood by few. I try to sleep but sleep eludes me.

And then the rooster crows again, the same four notes, this time a tremolo in the third. Almost imperceptibly, a grey light visits the eastern horizon. Silhouettes of palms emerge from their cloak of the night like images coming alive in a photographer's tray. A distant rooster now answers the call. Little birds, hidden in the trees, break into hesitant song. The light intensifies, shapes, concealed till now, fill the scene. Stars begin to fade at the approach of dawn.

A host of birds now joins in the chorus. Now and then one breaks into a delicate solo of exquisite but fleeting beauty. The sky in the east blushes with a hue of pink. Hitherto grey clouds turn to yellow, then burn into a deep orange. Birdsong intensifies. Doves, silent till now, engage in the chant.

And now the sky turns to gold; the sun is on its way. Billowing clouds dress in gleaming, festive white on its approach. A faint smell of smoke drifts on the air. Sounds of the awakening world visit the senses, the crescendo of a motorbike, a human voice, a baby's cry. Where before silence ruled, the hum of traffic now forms the backdrop to the new day, a thousand lives heading for unseen destinations, all intertwined for a moment in a symphony of sound. Birds now chatter in endless discourse, dart this way and that, driven by a purpose they alone know. A mother hen bids her brood to follow with a "bok-bok-bok-bok - today!" Dogs bark and a nearby cow bellows time and again, its resonant baritone forfeited on a single note.

Warm air embraces the landscape. A light breeze tempts the palms to a lazy, swaying dance. The stillness of the night is no more; there is movement all around.

At last the sun arrives in triumph. Flowers celebrate the coming of light, their reason for being, in an ecstasy of colour. Volcanoes of the island strut their proud profiles before retreating into their mantles of cloud. The stars, the darkness, the silence withdraw - let the sun rule the day. They know the ways of the sun, its flight across the sky. They bow to its role as the giver of life. They know its visit is passing. They know it will retire into the sea when its work is done.

And then the eternal ritual begins again.

Filed under: In Focus

One Response to "Bali Morning"

Michael Fibonacci Says:

May 28th, 2009 at 10:03 pm This is a beautiful piece of writing. It has a wonderful poetic quality, evoking images, sounds, colours in the ever changing melting pot which is the spirit of Bali awakening. I would love to know more about the author, Nicholas Rety, and read some more of his work.

Many thanks to The Bali Times for selecting such quality.

Michael Fibonacci

Copyright © 2008 The Bali Times - All Rights Reserved PT. Lestari Kala Media Jl. Oberoi, Seminyak, Bali, Indonesia Tel: 0361 737243 / 739407 Fax: 0361 739407 powered by WordPress | The Bali Times

EPILOGUE

These recollections are merely footprints, sundry events recalled through the mist of decades. A continuum about an ordinary life would sorely test the patience of the reader. Looking back I know that I was molded by my experiences and by the people with whom I shared the stage of life, like a stone shaped by the current of water in which it rests. All through my ups and downs I maintained a positive outlook and have remained comfortable with what I am. In times both good and bad I had the best seat in the house.

Regrets? I had some and I made my apologies where the fault was mine. Would I do things differently if I had a second crack at life? As my old friend told me, we do not have limitless options: we can only pick the best of the choices handed to us. That I tried to do, knowing that life is a one-way journey with no return ticket.

Nicholas Rety
Vernon, B.C., Summer 2021

CPSIA information can be obtained
at www.ICGtesting.com
Printed in the USA
BVHW040746200223
658843BV00005B/75